TVET TOWARDS INDUSTRIAL REVOLUTION 4.0

PROCEEDINGS OF THE TECHNICAL AND VOCATIONAL EDUCATION AND TRAINING INTERNATIONAL CONFERENCE (TVETIC 2018), JOHOR BAHRU, MALAYSIA, 26–27 NOVEMBER 2018

TVET Towards Industrial Revolution 4.0

Edited by

Nur Hazirah Noh@Seth, Norah Md Noor, Mahyuddin Arsat, Dayana Farzeeha Ali, Ahmad Nabil Md. Nasir, Nur Husna Abd. Wahid & Yusri Kamin

Universiti Teknologi Malaysia, Johor, Malaysia

Routledge
Taylor & Francis Group
LONDON AND NEW YORK

Routledge is an imprint of the Taylor & Francis Group, an informa business

Typeset by Integra Software Services Pvt. Ltd., Pondicherry, India

Published by: CRC Press/Balkema
Schipholweg 107C, 2316XC Leiden, The Netherlands
e-mail: Pub.NL@taylorandfrancis.com
www.crcpress.com – www.taylorandfrancis.com

ISBN: 978-0-367-24273-2 (Hbk)
ISBN: 978-0-429-28150-1 (eBook)
DOI: https://doi.org/10.1201/9780429281501

TVET Towards Industrial Revolution 4.0– Hazirah Noh@Seth et al. (eds)
© 2020 Taylor & Francis Group, London, ISBN 978-0-367-24273-2

Table of contents

TVET Towards Industrial Revolution 4.0– Hazirah Noh@Seth et al. (eds)
© 2020 Taylor & Francis Group, London, ISBN 978-0-367-24273-2

Preface

The TVETIC Proceedings contain a selection of papers accepted for presentation and scientific discourses at the Technical and Vocational Education & Training (TVET) International Conference 2018. The main theme of this conference was TVET Towards Industrial Revolution 4.0. The Scientific Committee for TVETIC 2018 was steered by the community of TVET practitioners, researchers and scholars from various disciplines and countries in the region. The committee had the responsibility to manage and evaluate (in a double-blind review process) the papers submitted for publication. The papers were clustered into the following subject areas: (a) Vocation and Learning, (b) Vocational Behavior, (c) Education 4.0, (d) Industry Revolution 4.0, (e) Innovative Leader, (f) Excellent talent, (g) Education transformation and change agent, (h) 21st Century Skills and Training, (i) Strategic Network, (j) Sustainability and green skills, (k) work based learning, (l) Green employability skill, (m) Assessment and Evaluation. TVETIC 2018 received 100 contributions from several countries in this area. The papers are highly significant to empower the quality education especially TVET in facing globalization and Industrial Revolution 4.0

Acknowledgements

On behalf of Universiti Teknologi Malaysia, we would like to thank all keynote speakers, presenters and participants at the 3rd Technical and Vocational Education & Training International Conference (3rd TVETIC 2018) which was held at KSL Hotel & Resort in Johor, Malaysia. We would also like to congratulate the Faculty of Social Sciences and Humanities, Universiti Teknologi Malaysia and Persatuan Pendidikan Teknik dan Vokasional Malaysia (PPTVM) for their great effort and success in planning and organising this conference. TVETIC is one of the platforms to build global linkage, collaboration and sharing of ideas in research to improve the quality of education, mainly on Technical and Vocational Education & Training (TVET) and related issues in education.

We would like to thank our distinguished keynote speakers, scientific committee, presenters and all participants for their contribution in making this conference a success. Special thanks and appreciation to our co-organisers: Universiti Pendidikan Sultan Idris (UPSI), BINUS University, Universitas Muhammadiyah, Palu, Indonesia, Rajamangala University of Technology Thanyaburi, Thailand, and Universitas Muhammadiyah Purwokerto, Indonesia for the tremendous support and commitment to this conference. Lastly, we would like to express our gratitude to the dedicated organising committee, members of the Faculty of Social Sciences and Humanities UTM and those who have directly and indirectly given their support for this conference. Thank you. Dr. Nur Husna Abd Wahid, Conference Chair of the 3rd TVETIC 2018.

TVET Towards Industrial Revolution 4.0– Hazirah Noh@Seth et al. (eds)
© 2020 Taylor & Francis Group, London, ISBN 978-0-367-24273-2

Committees

PATRON	Prof. Ir. Dr. Wahid Bin Omar
ADVISORS	Prof. Dr. Sukri Bin Saud Assoc. Prof. Dr. Yusri Bin Kamin
CONFERENCE CHAIR	Dr. Nur Husna Binti Abd. Wahid
CONFERENCE VICE CHAIR	Dr. Mohd Zolkifli Bin Abd. Hamid
SECRETARY	Dr. Rafeizah Binti Mohd Zulkifli
VICE SECRETARY	Dr. Zakiah Binti Mohamad Ashari
TREASURER	Dr. Hanifah Binti Jambari
VICE TREASURER	Dr. Mohd Rasidi Bin Pairan
FLOOR MANAGER	Dr. Mahyuddin Arsat
MANAGER OF PARALLEL SESSION	Assoc. Prof. Dr. Aede Hatib Musta'mal Dr. Nor Farahwahidah Binti Abdul Rahman
PUBLICITY/PROMOTION/WEBSITE/ APPS	Dr. Muhammad Khair Bin Noordin Mrs. Fadhilah Bte Othman Mrs. Azura Bte Baharudin
CERTIFICATES/SOUVENIR	Dr. Asha Hasnimy Binti Mohd Hashim Mrs. Roziah Bte. Tamin Mrs. Anisa Bte Ismail
ACCOMODATION AND MEAL	Dr. Nornazira Binti Suhairom Dr. Diyana Zulaika Binti Abdul Ghani
REGISTRATION/PROTOCOL/ INVITATION	Dr. Nurul Farhana Binti Jumaat Dr. Norulhuda Binti Ismail Dr. Nurul Aini Binti Mohd Ahyan
ARTICLES/PUBLICATION	Dr. Nur Hazirah Binti Noh@ Seth Dr. Ahmad Nabil Md Nasir Dr. Norah Bt Md Noor Dr. Dayana Farzeeha Binti Ali Dr. Mahyuddin Arsat
SPONSORSHIP	Assoc. Prof. Dr. Adnan Ahmad Assoc. Prof. Dr. Yusri Kamin Dr. Mohd Zolkifli Bin Abd. Hamid
PROGRAM BOOK	Dr. Muhammad Abd Hadi Bin Bunyamin
TRANSPORTATION AND TECHNICAL	Dr. Mohd. Fa'iz Bin Ahmad Mr. Ali B. Jawahir Mr. Shamsulrizal B. Abd Shukor
NETWORKING	Dr. Nor Fadila Binti Mohd Amin

SCIENTIFIC COMMITTEE

1. Dr. Nur Hazirah Binti Noh@ Seth
2. Dr. Ahmad Nabil Md Nasir
3. Dr. Norah Bt Md Noor
4. Dr. Dayana Farzeeha Binti Ali
5. Dr. Nur Husna Binti Abd. Wahid
6. Dr. Mahyuddin Arsat
7. Assoc. Prof. Dr. Yusri Bin Kamin

TVET Towards Industrial Revolution 4.0– Hazirah Noh@Seth et al. (eds)
© 2020 Taylor & Francis Group, London, ISBN 978-0-367-24273-2

TVET: Towards Industrial Revolution 4.0

R. Gennrich
Senior Expert TVET Governance, Germany

ABSTRACT: The quest for success in global competition in general – but in times of fast-changing global digitalization in production and life, in particular – incorporates the quest for highly efficient ways of optimizing the productivity and mobility of a nation's workforce. Consequently, national education and vocational training systems are called upon to be well prepared for rapid changes in life and production and to meet the new challenges of the future digital age. Training 4.0, Big Data, cyber-physical systems; this is the alphabet of buzzwords around the ongoing digital transformation in our societies. But, strictly speaking, they don't deliver concrete indications of how to deal with it. Business processes and forms of work, as well as social coexistence, are changing as a result of technological development and digitization.

1 THE CHANGING WORLD OF WORK

1.1 *"Winds of change"*

The world of work is changing rapidly and professions are changing and new ones emerging.

Networked production systems are often self-organizing. Their driving force is the processing of data and information, which specifies how machines are to be organized for a production order. The changes associated with digitalization affect the whole of society, the way we live and how we participate in social and political life. The future project, Industry 4.0, relates to an approach to industrial history, identifying four industrial revolutions. We are all contemporary witnesses at the beginning of the fourth Industrial Revolution. This new technological era, Industry 4.0, will have enormous consequences for education, training and employment. In the course of my following remarks I will use the term "Industry 4.0" instead of the even broader term of the fourth Industrial Revolution. I think that the talk of disruption through digitalization is too excited and inaccurate and I am convinced that we cannot overlook the implications of this development with regard to the future of work and life today.

1.2 *What does Industry 4.0 mean?*

Within the context of "Industry 4.0", it cannot be denied that work organization and work processes will change, along with proceeding automation and real-time oriented control. The same is true for work content and the interaction and communication between humans and technology, which have many consequences for educationalists as well as users and providers of the entire vocational training system. Industry 4.0 is the current trend of automation and data exchange in manufacturing technologies. It includes Cyber-Physical Systems (CPS), the Internet of Things (IoT), and cloud computing. In a broader sense, we must see Industry 4.0 as the next phase in the digitization of the manufacturing sector, driven by fundamental industrial transformations:

1. The astonishing rise in data volumes and connectivity, especially new low-power wide-area networks;
2. The emergence of analytics and business-intelligence capabilities;

3. New forms of human–machine interaction such as touch-based interfaces and augmented-reality systems;
4. Improvements in transferring digital instructions to the physical world, such as advanced robotics, smart maintenance and 3-D printing.

1.3 *10 target industries in the sphere of Industry 4.0*

The ten target industries in the focus of almost all governments in order to respond to the rapidly changing world of work and life in the digital age are:

1. Next-generation automobiles (electric cars)
2. Smart electronics
3. Robotics
4. Affluent medical and wellness tourism
5. Agriculture and agricultural processing industries
6. Biotechnology
7. Food processing (ready-to-eat and food for the future, also called "digital food")
8. Aviation and logistics
9. Biofuels and biochemicals
10. Urbanization connected with digitalization.

All of the targeted industries will focus on added value through advanced technology and innovation, which means that all ASEAN countries, including Malaysia, need investment in those areas.

1.4 *Digitalization in production and life*

Digitalization is not only advancing in the industrial sectors! Industry 4.0 is also marked by technological breakthroughs in the services and banking sectors among others. Let's take a look at the fast-growing urbanization. The urban infrastructure consists not only of glass and concrete, but also of bits and bytes. Digitalization therefore shapes apartments, business premises and open-plan offices, as well as urban transport (e.g. the cloud-based, open IoT operating system "MindSphere" from Siemens helps smart cities to network infrastructures, buildings and traffic). This example shows how profound the changes are that stand before us. Education experts and government agencies need to question whether our Technical and Vocational Education and Training (TVET) concepts and programs will meet future needs and what we need to change.

1.5 *Moving with the "future"*

A robust study by the University of Bremen, Germany, on "Industry 4.0 – Impacts on Initial and Further Training in the Metal and Electrical Industry", has shown that software technical networking and its related CPS elements are continuously becoming more prevalent. One guiding question of the study was: How far have companies advanced with the implementation of Industry 4.0 and what should be done by TVET to respond to the needs and expectations of the new technological challenges?

The challenges resulting from the implementation of Industry 4.0 can be summarized as follows:

1. Skilled workers, technicians, high technicians as well as vocational teachers and trainers should be qualified for specializations relevant to Industry 4.0. They must be able to master processes in their complexity and to safeguard the flawless operation of plants.
2. The mastering of networked systems with decentralized intelligence, the ability to deal with data and its analysis count among the most important requirements for work on production sites.

3. The thus-far setting of priorities on general questions of Industry 4.0 has to be extended by technological priorities (CPS), and by issues of work organization.

These trends show that skilled staff with high-quality initial and/or further training oriented to Industry 4.0 will probably have very good employment and career opportunities in the future. This finding is contradicted by some experts who state that especially skilled workers will be without a chance, or even at risk of extinction, due to the developments towards Industry 4.0. We are convinced that qualified personnel are needed as driving forces as in any other periods of technological revolution.

1.6 *Conclusion*

Digital technology is invading the world of work. It's not just physical work that has been lost but also cognitive work is no longer considered to be exclusive to human beings. The future of work is promising but it calls for greater care than ever. A final statement about companies' readiness and the full extent of skills needs in relation to Industry 4.0 across all sectors is currently not yet possible. But the above-mentioned survey shows changes related to the implementation of Industry 4.0 and states clearly that there will be a change of paradigms for skilled workers and technicians in production. Those changes have far-reaching consequences in relation to the philosophy and the entire system of vocational education and training.

2 INEVITABLE CHANGES IN THE WORLD OF EDUCATION AND TRAINING

The next production revolution, with global connectivity and smart technologies, is rewriting the concept of jobs and skills for the future, and consequently affecting learning and teaching experiences for TVET teachers and learners.

2.1 *New professional requirements resulting from Industry 4.0*

Table 1 provides an overview of the new occupational competencies resulting from digitalization and shows how technological change leads to newly required professional competencies.

2.2 *Top ten skills in the context of Industry 4.0*

According to a Future and Jobs Report of the World Economic Forum from 2015, the top ten skills of relevance in the context of Industry 4.0 are: (1) complex problem-solving, (2) critical thinking, (3) creativity, (4) people management, (5) coordinating with others, (6) emotional intelligence, (7) judgment and decision-making, (8) service orientation, (9) negotiation, and (10) cognitive flexibility. I would like to draw your attention to the increasing importance of cross-functional skills. Those skills (also known as soft skills) are interpersonal skills that are not job-specific. As I said before, those competencies and abilities are not new but still valid and are becoming more important. We should ask ourselves how these demands are being met and what remains to be done so that vocational education and training can better meet the needs of Industry 4.0.

2.3 *Top challenges on key levels of intervention*

Shaping TVET for the future is a complex task that involves the entire TVET structure and people on different levels of intervention:

1. Policy level;
2. Institutional level;
3. Implementation level.

The key topics show that TVET is lagging behind the challenges and requirements of Industry 4.0 in many respects and in many countries.

Table 1. Newly required competencies deriving from technological changes in production.

Technological production systems	Newly required competencies (examples)
▪ Cyber-physical systems ▪ Connectivity (Internet of Things) ▪ New configurations in production ▪ Cloud engineering ▪ Interactive communication platforms ▪ Smart workstations ▪ RFID[1] in production	✓ Understand the complete manufacturing process (starting from the order to the delivery of the final product) ✓ Work with smart devices on the shop-floor level ✓ Communication with and handling of robots (e.g. smart-connected robotics, smart maintenance) ✓ Understand how to receive and to visualize data in real time ✓ Using and programming firm software ✓ Work with data mining and with clouds

3 CONCLUSIONS AND RECOMMENDATIONS

3.1 *Vocational teacher education (Pedagogical aspects)*

Shifting Vocational Teacher Education (VTE) towards the new trends of Industry 4.0 means, among other things, accepting that a new culture of teaching and learning in TVET will be necessary! Vocational pedagogy and didactical principles need to be reassessed and be newly defined. Therefore, please allow me to draw your attention to the progressing position of vocational learners in the process of learning and working:

(1) Vocational learners will have more opportunities to learn at different times in different places. eLearning tools facilitate opportunities for remote, self-paced learning. Classrooms will be flipped, which means the theoretical part is learned outside the classroom, whereas the practical part shall be taught face to face, interactively.

(2) Vocational learners will learn with study tools that adapt to the capabilities of the student. This means above-average students shall be challenged with harder tasks and questions when a certain level is achieved. Students who experience difficulties with a subject will get the opportunity to practice more until they reach the required level.

(3) Vocational learners will adapt to project-based learning and working tasks. This means they have to learn how to apply their skills in shorter terms to a variety of situations.

(4) Schools, together with companies, will provide more opportunities for vocational learners to obtain real-world skills that are representative of their jobs. This means learning approaches like work-process-based learning (in Germany, DCT) curricula will create more room for students to fulfill internships, mentoring projects and collaboration projects, for example.

(5) The human interpretation of data will become a much more important part of the future curricula. Applying theoretical knowledge to numbers, and using human reasoning to infer logic and trends from these data will become fundamental new aspects of this literacy (STEM).

(6) Assessment and testing systems in TVET will be changed. Because the factual knowledge of a student can be measured during their learning process, the application of their knowledge is best tested when they work on projects in the field.

1. Radio Frequency Identification (RFID) has started to exert a major influence on modern supply chain management. In manufacturing, RFID changes the way objects are tracked on the shop floor and how manufactured goods interact with the production environment.

(7) Students will incorporate more independence into their learning process, such that mentoring will become the fundamental teaching approach. Teachers will form a central point in the jungle of data and information for students.

These are exciting, provocative and potentially far-reaching challenges. For individuals and society, new educational tools and resources hold the promise of empowering individuals to develop a fuller array of competencies, skills and knowledge, and of unleashing their creative potential.

3.2 *Vocational teacher education (Technical and didactical aspects)*

With regard to the ongoing changes, TVET experts and practitioners in VTE and teacher training institutions must intensify their efforts in order to revise teaching contents – in both directions – for TVET and VTE. Referring to a study of the University of Bremen, Germany, nine generic occupational fields of action relevant for Industry 4.0 were generated from the empirical fields of action in German companies in the metal and electrical industry. These "generic" occupational fields of action are exemplary of the entire transformation process and emphasize "the new" relevant contents for metal and electrical occupations.

Their target expectations are as follows: (1) simulation of plants; (2) plant networking; (3) ensuring the data availability from sensors; (4) monitoring, analysis, and evaluation of real-time data; (5) guaranteeing process safety through process monitoring and repair; (6) safeguarding machine data for plant quality operation, analyzing operational data and optimizing processes; (7) preventive, foresighted maintenance, multi-functional machines, assessment and use of different data and data formats; (8) considering repair interdependencies due to networking and IT-integration of machines and plants, including software updates; (9) diagnosis and troubleshooting in networked plants.

These generic occupational fields refer to the German context but, in a globalized context of Industry 4.0, similar reflections and conclusions shall be relevant for the adaptation of new training contents with regard to the same industrial sectors in Malaysia and other ASEAN countries.

3.3 *Vocational teacher education (Cross-cutting pedagogical aspects)*

The high complexity of occupational skills and competencies for global competitiveness in the 21st century shows an increasing importance for so-called "core soft skills". Such competencies can only be promoted decisively if appropriate support systems are developed and made available at the same time. The core subjects of 21st-century themes are composed as follows:

1. Life and career skills;
2. Learning and innovation skills;
3. Information, media and technology skills.

Furthermore, language competence (English language) becomes more and more relevant in high-tech occupations and should therefore be introduced into TVET as obligatory.

3.4 *Vocational teacher education (Priority measures)*

Taking the above-mentioned remarks (explanations) into account, the following priorities are urgently needed in order to improve vocational teacher education:

1. **Action-oriented research is being given a high priority**. In the age of digitalization, the challenges facing vocational school teacher training cannot be mastered without research. The best way to do this will be to promote research projects in conjunction with specialists and experts from companies, researchers and students;
2. **VTE structures, contents and methodologies must be updated fundamentally.**
3. **The teacher's job profile changes fundamentally**: teachers and trainers will be more mediator than teacher or trainer! Problem-based learning, a flipped classroom, and curriculum

design can be tools to help improve learning outcomes and better prepare people for continuous learning in the Industry 4.0 world of work.

4. **Close cross-regional (research and) cooperation** between universities, training institutions and companies is an essential prerequisite for successfully mastering the challenges posed by Industry 4.0. Therefore, all initiatives pursuing this goal should be actively supported by governments, international organizations and donors. One particular example of cross-regional and cross-institutional cooperation and experience sharing is RAVTE (see below).

3.5 *Vocational teacher education – One priority for the future*

VTE is among the most underestimated but most efficient core elements for gaining advantage in TVET quality improvement! Mastering the current and future challenges of Industry 4.0 without well-performing TVET personnel is an illusion. This challenge has been the reason for bringing the best experts from leading VTE institutions in Asia under one roof – the Regional Association for Vocational & Technical Education (RAVTE). RAVTE is an exclusive and independent regional association in Asia focusing on ASEAN integration and harmonization in the TVET and Skills Development (SD) sectors. According to the principles and goals of the regional cooperation network, RAVTE: "Vocational teachers are the transmitters of any kind of reform measures and challenges, and have to be well prepared for the current and new challenges of the 21st century".

3.6 *Recommendations on TVET policy level – Final remarks*

In response to the emerging requirements, TVET needs a well-performing *Vocational Education and Training Policy & Strategy for Industry 4.0*. Governments, assisted by leading companies, should provide reliable and sufficient financial resources to modernize TVET with regard to Industry 4.0 and the future qualification needs of our young generation. The TVET–Industry 4.0 policy should respect the global perspective but must be equipped with customized solutions and implementation mechanisms at regional and national levels. From my point of view, key reform measures of an "Industry 4.0 Policy" that can help to move TVET into the future should include, as a minimum:

1. Future-oriented legislation with regard to Industry 4.0;
2. Technological upgrading and modernization of VTE systems and facilities;
3. Application of workplace-based training and cooperation with the business sector (increasing importance of Work Integrated Learning (WIL) and Dual Training);
4. Action-oriented and interconnected TVET research (and decision-making);
5. VTE strategy for Industry 4.0;
6. Reliable and sufficient financial support.

I'm very pleased to see that this important topic has brought together so many interested experts here at *Universiti Teknologi Malaysia* (UTM) Johor Bahru, and I'm convinced that most of you belong to the optimistic-minded group of people that are proactively involved in your daily work as lecturer and researcher, teacher or practitioner in order to shape the future of TVET in your workplaces. I am very grateful to the organizers for this invitation to speak and hope that I have been able to contribute to a successful outcome for the conference with my modest reflections about the "Future Challenges of TVET".

Industry 4.0 is the driving force of the fourth Industrial Revolution and should be seen by all of us working in governments, universities, companies and vocational training institutions as a chance to reform TVET and to provide a better future for our young generation.

TVET Towards Industrial Revolution 4.0– Hazirah Noh@Seth et al. (eds)
© 2020 Taylor & Francis Group, London, ISBN 978-0-367-24273-2

The strategic collaboration formation phase between vocational colleges and automotive industries in Malaysia

Norisham Bin A. Rahim, Yusri Bin Kamin & Halliru Shuaibu
School of Education, Faculty of Social Sciences & Humanities, Universiti Teknologi Malaysia, Malaysia

ABSTRACT: The purpose of this study was to investigate the development phase in strategic collaboration between educational training institutions and automotive industries. In line with the aspiration to become a high-income developed country, besides increasing economic complexity and technological advances, some efforts need to be made to provide quality skilled workers to meet industry requirements. The objective of this study was to identify the constructs in the development phase in the formation of strategic collaboration between vocational colleges and automotive industries in Malaysia. An intrinsic case study, a type of qualitative research approach, was used as the design of the study. The study used multiple instruments, namely document analysis and semi-structured interviews, in order to develop an insight into stakeholders' perspectives on the strategic collaboration formation phase between vocational colleges and automotive industries. Documents analyzed in this study included the memorandum of understanding, industry statements and a letter of intent/tour of intent signed between the vocational colleges and the industry. Twenty automotive lecturers were drawn from vocational colleges to make the sample of the study. All 20 automotive lecturers participated in the semi-structured interviews as respondents. The findings identified the development phase in the formation of collaboration between vocational colleges and automotive industries. The research also found three phases in collaboration undertaken by the vocational colleges such as a before implementation phase, an implementation phase and an after implementation phase.

1 INTRODUCTION`

Collaboration is a complex relationship as it requires a solid source (Buang, 2016). With the resources for and commitment to the establishment of collaboration, it will be smoother and easier to achieve goals and objectives. Collaboration between educational institutions and industries is a field that is indispensable in developing skills and knowledge in the world of work and education (Guimon, 2013). Therefore, to implement it, the categories, requirements and methods appropriate to the objectives and goals to be realized should be known. The findings of a previous study, conducted by Patel and D'Este (2007), also explains that there are five strategies that can be applied in creating collaboration between educational training institutions and industries, first through official meetings and conferences, consultancy and research, sharing technology and the fifth is the provision of training and carrying out research work in concert. According to Chin (2011), collaboration between educational training institutions and industries is classified into two types, namely formal and informal collaboration. Researchers have given explanations regarding the implementation of this collaboration, which covers both categories, formal and informal. Formal collaboration is a method of cooperation, teamwork, partnership or alliance through a formal approach and adherence to procedures and the consent of both sides. Hagedoorn et al. (2000) mentioned that creating a formal collaboration involves activities such as contract deals, research projects, licensing of product patents, publications, exchange of labor and expertise, conferences and so on. The period for implementation can be done either in the short- or long-term depending on the needs and situation (Guimon, 2013). Formal collaboration carried out over a long time, according to Koschatzky and Stahlecker (2010), is a strategic collaboration because it provides an opportunity for both sides to build methods and evaluate the ability to implement a collaboration, which in turn can solve the problems or issues more effectively.

1.1 *Vocational colleges*

Malaysia needs a highly skilled workforce to support the growth of the industrial sector, which has led to the establishment of many vocational and technical training institutes in the fields of engineering, business and services. Vocational schools were revamped in the transformation of Technical and Vocational Education and Training (TVET) to become vocational colleges that aim to produce skilled workers for industries. Vocational education is offered by educational institutions under the Ministry of Education (MOE), which includes vocational colleges, community colleges, polytechnics and the Malaysian Technical University Network (MTUN) universities. The qualifications offered by vocational colleges include certificates and diploma.

Vocational colleges offer various courses such as courses in the fields of electrical and electronics engineering technology, mechanical and manufacturing engineering technology, civil engineering technology, hospitality and tourism, field of study and public services, business, information technology and communication and agriculture. The objective of the Vocational College Program is to produce a skilled and competent workforce to meet the needs of industry and entrepreneurs. One of the most demanding programs is the Automotive Technology Program. This program provides basic vehicle engine exposure to the interest of students to prepare and strive to extend learning to the highest level in the automotive field, being skilled in human capital and contributing toward national development. In order to produce competent and skilled students, the facilities and tools used for the learning and teaching process should be in line with current technology. The development of technology, especially in the automotive field, is seen to be rapidly evolving from conventional technology to current technology like Natural Gas Vehicles (NGVs), hybrid and electric technologies. So far, there are nearly 50 vocational colleges in Malaysia that offer automotive technology programs.

1.2 *Automotive industry*

The creation of the automotive industry in Malaysia has corresponding needs in terms of skilled workers and reliance on creativity and innovation from a skilled workforce. Some employers argue that graduates from higher-education programs do not meet the industry's needs. There are needs for collaboration from the automotive industry to help the education system especially in TVET. The importance of automotive engineering is to provide the country with a step up to the next level by producing automotive technology. According to Kamin (2012, p. 14), involvement in the production of automobiles can generate economic growth, in particular employment for engineers, team leaders, technicians, skilled tradespeople and semi-skilled operators on production lines. He added that the industry creates many opportunities for local talent to be involved in the automotive industry rather than depending on foreign expertise. The development of the automotive industry encourages the workforce to be innovative and creative in order to be globally competitive.

An increase in the amount of hands-on human capital in the national workforce is needed to fulfill the expansion of industries in Malaysia. The Malaysian Employers Federation (2004) states that there is a declining number of graduates in science and technical programs. Another major concern is unemployment due to graduates lacking some skills required in the workplace. For example, employers expect someone who is 'work ready'; they do not want to spend money and time retraining new graduates in the workplace environment.

2 LITERATURE REVIEW

Tracer studies conducted by researchers indicate that phase formation and collaboration were explored using a variety of methods. Thus, through the establishment of phase highlights, a diversity of collaborative writing is described as a reference and guide to researchers across the span of 17 years, starting from early 2001 until 2017 (2001–2017). Therefore, the researchers will analyze relevant meta-phase formation as well as describe the relevant details of the meta-analysis illustrated in Table 1.

Table 1. Meta-analysis of the collaboration formation phase.

No.	Author(s)	Activities	Phase
1	Dyer, Kale and Sigh (2001)	• Pre-review and evaluation • Selection • Discussion and negotiation • Collaboration management • Evaluation	• Before implementation (intra-organization) • Implementation (interorganization and cooperation) • After implementation (review and outcomes)
2	Pett and Dibrell (2001)	• Agenda discussion • Needs analysis • Risk evaluation • Sharing information • Agreement • Reliability assessment • Cooperation • Effectiveness evaluation and potential	• Before implementation (exploratory stage) • Implementation (recurrent and rational contract stage) • After implementation (outcome stage)
3	Lendrum (2003)	• Selection • Needs analysis • Site visit • Technology needs analysis • Build trust • Negotiation • Implementation strategy • Evaluation	• Before implementation (workplace reform and capability) • Implementation (partnering process) • After implementation (quality of collaboration)
4	Wahyuni (2003)	• Goal and needs setting • Formation • Selection • Agreement • Implementation (conflict/communication/control) • Collaboration performance • Evaluation	• Before implementation (formation) • Implementation (operation) • After implementation (evaluation)
5	Cools and Roos (2005)	• Strategy compilation • Searching • Discussion and negotiation • Collaboration management • Performance evaluation • Reporting	• Before implementation (intra-organization) • Implementation (interorganization and cooperation) • After implementation (review and outcomes)

(*Continued*)

Table 1. (*Continued*)

No.	Author(s)	Activities	Phase
6	Shenkar and Reuer (2006)	• Preparation • Opportunities identification • Objectives review • Collaboration design • Implementation • Evaluation performance	• Pre-formation • Formation • Post formation
7	Schanan and Kelly (2007)	• Rationalization • Negotiation • Implementation	• Pre-implementation • Implementation
8	Tjemkes, Vos and Burger (2012)	• Selection • Strategy formulation • Negotiation and discussion • Collaboration design • Management and implementation • Collaboration evaluation	• Before implementation (inter- and intra-organizational phase) • Implementation (cooperation phase) • After implementation (evaluation and termination phase)
9	Bhandari and Verma (2013)	• Searching • Selection • Planning • Objective and goal setting • Agreement • Evaluation	• Review phase • Formation strategy phase • Implementation strategy phase • Strategy evaluation and control phase
10	Nevin (2014)	• Selection criteria development • Evaluation • Preferred partner shortlist • Mapping • Documentation • Develop vision and mission • Support system • Evaluation	• Before implementation (searching and analyzing) • Implementation (developing plan) • After implementation (evaluation and supporting tools)
11	Elvers and Song (2014)	• Searching • Selection • Collaboration matrix development • Implementation • Evaluation	• Before implementation (data collection) • Implementation (analysis network) • After implementation (correlation between strategic partnering behavior)
12	Ankrah and Al Tabbaa (2015)	• Searching • Interaction • Evaluation and selection • Discussion • Agreement	• Formation process phase
13	Muthoka and Kilika (2016)	• Selection • Vision and mission setting • Cost and sources analysis • Collaboration classification • Implementation	• Before implementation (partner selection) • Implementation (formation)
14	Russo and Cesarani (2017)	• Needs analysis • Selection • Implementation strategy	• Before implementation (formation phase)

(*Continued*)

Table 1. (*Continued*)

No.	Author(s)	Activities	Phase
		• Coordination and management • Discussion and promotion • Agreement • Performance evaluation	• Implementation (operational phase) • After implementation (evaluation phase)
15	Gebrekidan (2017)	• Planning • Types of collaboration determination • Selection • Objective and strategy discussion • Evaluation • Periodic review	• Before implementation (pre-strategic alliance phase) • Implementation (strategic alliance phase) • After implementation (post-strategic alliance phase)

Based on the meta-analysis of the collaboration formation phase performed in Table 1, it could be safely and generally declared that three phases are involved, namely: a phase prior to the implementation; the phase of implementation; and a post-implementation phase. All three phases have different implementation activities.

3 STATEMENT OF PROBLEM

There are several key factors that led to a collaboration that is designed not to be executed properly. Factors in selecting a collaboration partner are the main ones to be considered. Most previous studies indicated that the selection of the right collaboration partners contributes to the preservation of an established collaboration (Chin, 2011; Barry & Fenton, 2013; Kauppila et al., 2015: Filippetti & Savona, 2017). This is because of compatibility and understanding of the situation and collaboration partner institutions and industry conditions, which are very important because they contribute to generation of an effective working culture. Based on studies and the literature references conducted, it has been proven that differences in interests and expectations, as well as work culture, between the two types of organization are factors in the establishment of planned collaboration.

Cultural differences that exist may be due to the two organizations having different objectives and goals, the differences in orientation and scope of the task, and also significant differences in the characteristics of the two organizations (Thune, 2011; Iqbal, 2013; Kapil, 2014; Malik & Wickramasinghe, 2015; Calcagnini et al., 2015; Lemos & Cario, 2017). The margin of difference usually leads to conflicts between the interests and expectations of the future direction of the established collaboration (Chin, 2011; Rashidi, 2013). This is the case because one party is more educational in nature, focusing on providing the skills and knowledge for the industry-focused culture, while the other is working efficiently to contribute to increasing profits and investment. The difference in terms of objectives, goals, orientation and characteristics of the organizations mostly lead to the failure of the formation of a collaboration and cause industries to fail to recognize educational institutions as good collaboration partners (Philbin, 2008; Malik & Wickramasinghe, 2015; Taratukhin et al., 2016). This situation is also closely linked to the issue of trust and confidence; as elements of confidence and trust almost always help to build an understanding of the objectives and needs of the organizations involved (Bruneel et al., 2010; Nevin, 2014; Filippetti & Savona, 2017; Lemos & Cario, 2017).

The researchers also carried out a preliminary study through an analysis of documents, involving the memorandum of understanding, industry statements and letter of intent signed between the vocational colleges and the industries in the collaboration. The results of the document analysis revealed a number of issues such as that the memorandum of understanding, industry statements and letters/tours of intent mainly focused on training the trainers and lecturer placement only. Policy documents skewed emphasis on quality of collaboration as well as factors for choosing and forming effective collaboration that would yield the desired goal for vocational colleges and industry alliances. This clearly shows a gap that needs to be closed so that not only would collaboration be formed to meet Key Performance Indicators (KPIs) at colleges but also existing collaborations should be used properly in order to improve the quality of the vocational college education system. Certainly, effective collaboration will create a sound grounding for vocational college graduates in the industrial sector and they would become competitive at the national and international levels.

It is obvious that collaboration between industry and educational training institutions in Malaysia, especially in vocational colleges, is still poor and the ground it covers is narrow. In addition, the framework of existing collaboration emphasizes strategy implementation and places less emphasis on elements of administration and evaluation. Mechanisms and an appropriate collaboration framework for addressing these issues need to be established so that the quality of education and training systems can meet the needs of the industry, which in turn can increase productivity. Requirements related to the study of collaboration between educational training institutions and industries remain relevant because the current rapid pace of technological change encourages all parties involved in forming collaborations to ensure that the graduates produced by vocational colleges be highly skilled to join the workforce and meet the needs of the job market. Although a few studies have been carried out abroad on collaboration between educational institutions and industries, the results of these studies may not be suitable for use in Malaysia due to differences in the culture and, the education system.

4 OBJECTIVES OF THE STUDY

The objective of this study is to identify the main constructs and subconstructs in the formation phase of strategic collaboration between vocational colleges and the automotive industries in Malaysia. To achieve this objective, the following research questions were formulated to guide the study:

i. What are the main constructs and subconstructs of the strategic collaboration formation phase according to the document analysis?
ii. What are the main constructs of the strategic collaboration formation phase from the perceptions of vocational college lecturers?

5 METHODOLOGY

The research methodology describes the procedures followed in conducting the study, which are discussed in the following subsections.

5.1 *Research design*

The research design is the blueprint or conceptual structure for collection, measurement and analysis of data obtained from a study. It also refers to the methods and procedures used to collect detailed information and give a clear picture so as to position the findings and conclusions of a study (Gupta, 2005; Abbott & Bordens, 2011). In this study, an intrinsic case study, which is a type of qualitative research design, is used to obtain information and to see the implementation of the strategic collaboration between vocational colleges and the automotive industries in Malaysia.

5.2 *Area of the study*

Vocational colleges in Johor, Melaka, Negeri Sembilan, Pahang and Kedah were purposely selected as the area of the study and the reason for their selection is that they were among the oldest and best-stocked vocational colleges that specialize in the automotive training of students in Malaysia.

5.3 *Sample and sampling techniques*

The sample population of the present study is shown in Table 2. Twenty vocational college lecturers were purposively selected as the sample of the study. Vocational college lecturers in the context of this study refer to teachers of automobile programs who work in vocational colleges and were selected due to their rich experiences in the field. The choice of 20 participants is in line with the recommended sample size, ranging from five to 25, for a phenomenological study (Creswell, 2013). Moreover, purposive sampling is viewed as judgmental sampling to represent the population of interest without sampling at random, deciding what must be known by an individual or groups, and setting out to find persons available and willing to participate in a study by providing information based on their knowledge or experience (Collins et al., 2007).

Table 2. Tabulation of respondents' interviews.

No.	Study area	Vocational colleges	Respondents
1	Johor	Vocational College A	4
2	Melaka	Vocational College B	4
3	Negeri Sembilan	Vocational College C	4
4	Pahang	Vocational College D	4
5	Kedah	Vocational College E	4
	Total		**20**

5.4 *Instrumentation*

According to Creswell (2009), document analysis involves researchers examining documents such as minutes of meetings, newspaper articles and other related documents in order to understand the issues and problems under investigation. Document analysis also means a process of gathering information through the exploration and analysis of written materials appropriate to a particular study. Accessible documents are categorized into two: shared documents and confidential documents. The researchers conducted the document analysis by gathering, examining, analyzing and synthesizing data from a memorandum of understanding involving vocational colleges and automotive industries, industry statements and letters/tours of intent. In addition, the researchers used semi-structured interviews; a method of individual interviewing to obtain information from vocational college lecturers. According to Creswell (2007), interviews conducted individually are more ideal as they provide a comfortable space and opportunity to share ideas without feeling hesitant to express an opinion and to explain the issues clearly. Similarly, Sidek (2002) reported that individual interviews allow the information obtained to be more personal. A semi-structured interview protocol was developed by the researchers and validated for clarity of statement, construct and content coverage by two measurement and evaluation experts. Member checking was used to establish the reliability of the interview protocol. Member checking simply means giving the instrument back to the participants after the interview for confirmation of item statements captured during the interview session. The permission of participants was obtained to record the interview session for the purpose of analysis. ATLAS.ti 8 analysis software was used for transcription, thematic development and coding of the data obtained from the semi-structured interviews.

5.5 *Scope and limitations of the study*

In general, this study is limited to the determination of the main constructs and subconstructs of the information phase in the strategic collaboration between vocational colleges and the automotive industries in Malaysia through the use of document analysis and semi-structured interviews. The study focused on 20 vocational college lecturers in the field of automotive technology in vocational colleges under the Ministry of Education, Malaysia. Five vocational colleges were involved in the study from Johor, Melaka, Negeri Sembilan, Kedah and Pahang.

6 ANALYSIS AND DISCUSSION

Based on the objectives of this study, the following results were obtained from the document analysis (as presented in Table 3) and the semi-structured interviews (as presented in Table 4).

Table 3. Periodical review (summary of document analysis findings).

		Before					Implementation					After	
Author(s)	Year	B1	B2	B3	B4	B5	I1	I2	I3	I4	I5	A1	A2
1.Dyer, Kale and Sigh	2001			/		/		/				/	/
2.Pett and Dibrell	2001		/	/				/	/		/	/	/
3.Lendrum	2003		/	/		/	/	/	/	/	/		/
4.Wahyuni	2003	/	/	/		/	/	/	/	/	/		/
5.Cools and Roos	2005	/		/			/	/	/			/	/
6.Shenkar and Reuer	2006	/	/	/		/	/	/	/	/	/	/	/
7.Schanan and Kelly	2007		/		/		/	/	/	/	/		/
8.Tjemkes, Vos and Burger	2012					/	/	/	/	/	/	/	/
9.Bhandari and Verma	2013	/		/		/	/				/		/
10.Nevin	2014		/	/	/	/	/	/	/		/	/	/
11.Elvers and Song	2014		/	/	/		/	/	/	/	/		/
12.Ankrah and Al Tabbaa	2015			/			/	/	/				/
13.Muthoka and Kilika	2016	/	/			/	/	/	/	/	/		/
14.Russo and Cesarani	2017	/	/			/		/	/		/	/	/
15.Gebrekidan	2017	/	/			/	/	/	/				/
TOTAL		7	10	10	3	10	12	14	13	7	1	7	12

6.1 *Main constructs and subconstructs of the strategic collaboration formation phase from the document analysis*

Table 3 shows the results of the research analysis related to the collaboration formation phase undertaken by previous researchers. It is derived from the analysis of literature materials relating to forming the collaboration phase between industry and educational institutions and training from the start of 2000 to now.

B1	Planning	I1	Preparation	A1	Collaboration management
B2	Needs analysis	**I2**	Discussion	**A2**	Evaluation
B3	Searching	**I3**	Negotiation		
B4	Pre-review	**I4**	Site visit		
B5	Selection	**I5**	Agreement		

6.2 *Main constructs of the strategic collaboration formation phase from the perception of vocational college lecturers*

Table 4 shows a summary of the findings of the interviews that were conducted with vocational college lecturers in the field of automotive technology. From the interviews, the establishment phase of collaboration between vocational colleges and the automotive industry in Malaysia shows that the three phases applied were: before the implementation phase; the implementation phase; after the implementation phase. A number of important activities were discovered in this study which were found to be similar to the findings of previous researchers, although other researchers did not execute their research at vocational colleges. For example, pre-review of the industries selected for the study. It is important to ensure that the industries selected are appropriate and can assist with vocational colleges' requirements. In addition, needs analysis: the activity of identifying the real needs of vocational colleges to establish collaboration was usually not done in previous studies. Needs analysis is crucial because it helps the colleges understand their situation and the needs for establishing collaboration.

Table 4. Semi-structured interview results.

	Before					Implementation					After	
Respondent	B1	B2	B3	B4	B5	I1	I2	I3	I4	I5	A1	A2
R1	/	/	/		/	/	/	/		/	/	
R2	/		/		/	/	/	/	/	/		
R3	/	/	/	/	/	/	/	/	/	/	/	
R4	/	/	/		/	/	/	/	/	/		
R5	/	/	/	/	/	/	/	/	/	/	/	
R6	/		/		/	/	/	/	/	/		
R7	/		/	/	/	/	/	/	/	/	/	
R8	/	/	/		/	/	/	/		/	/	
R9	/	/	/	/	/	/	/	/		/	/	
R10	/		/	/	/	/	/	/	/	/	/	
R11	/	/	/	/	/	/	/	/		/	/	
R12	/		/		/	/	/	/	/	/	/	
R13	/	/	/	/	/	/	/	/	/	/	/	
R14	/		/		/	/	/	/		/		
R15	/	/	/	/	/	/	/	/	/	/	/	
R16	/		/		/	/	/	/		/	/	
R17	/	/	/		/	/	/	/	/	/		
R18	/		/		/	/	/	/		/	/	
R19	/		/	/	/	/	/	/	/	/	/	
R20	/		/		/	/	/	/	/	/	/	
Total	20/20	10/20	20/20	9/20	20/20	20/20	20/20	20/20	13/20	20/20	15/20	0/20

B1	Planning	I1	Preparation	A1	Collaboration management
B2	Needs analysis	**I2**	Discussion	**A2**	Evaluation
B3	Searching	**I3**	Negotiation		
B4	Pre-review	**I4**	Site visit		
B5	Selection	**I5**	Agreement		

7 CONCLUSION

In conclusion, the collaboration formation phase has three phases which are: the phase before implementation; the phase of implementation; the phase after execution. Each phase has appropriate activities and can ensure the formation of collaboration to achieve its objectives. It can indirectly create an effective and more organized collaboration. The uniqueness of this study is that it was conducted at vocational college level with the hope that its findings will inform policymakers in improving collaboration between vocational colleges and automotive industries in Malaysia. Hence, the study is a contribution to the associated body of knowledge.

ACKNOWLEDGMENTS

The authors would like to thank the Ministry of Higher Education and *Universiti Teknologi Malaysia* (UTM) for Research University Grant (Tier 1) that was used to support this research. Vote grant PY/2017/01682.

REFERENCES

Abbott, B.B. & Bordens, K.S. (2011). *Research design and methods: A process approach* (8th ed.). Boston, MA: McGraw-Hill.

Ankrah, S. & Al Tabbaa, O. (2015). University–industry collaboration: A systematic review. *Journal of Management, 31*, 387–408.

Barry, A.M. & Fenton, M (2013). University–industry links in R&D and consultancy in Ireland's indigenous high-tech sector. *Irish Geography, 46*, 51–77.

Bhandari, A. & Verma, R.P. (2013). *Strategic management: A conceptual framework*. New Delhi, India: McGraw-Hill Education.

Bruneel, J., D'Este P., dan Salter, A. (2010) *Investigating the factors that diminishthe barriers to university–industry collaboration*. Research Policy 39 (2010) 858–868

Buang, Z.B. (2016). Collaborative Relationships In The Process Of Helping A Skilled Community Through Lifelong Learning Johor, Malaysia: Ledang Community College

Calcagnini, G., Liberati, P., Giombini, G. & Travaglini, G. (2015). A matching model of university–industry collaborations. *Small Business Economics, 46*, 31–43.

Chin, C.M.M. (2011). *Development of a project management methodology for use in a university–industry collaborative research environment* (Doctoral thesis, University of Nottingham). Retrieved from http://eprints.nottingham.ac.uk/12941/.

Collins, K.M.T., Onwuegbuzie, A.J. & Jiao, Q.G. (2007). A mixed methods investigation of mixed methods sampling designs in social and health science research. *Journal of Mixed Methods Research, 1* (3), 267–294. doi:10.1177/1558689807299526.

Cools, K. & Roos, A. (2005). *The role of alliances in corporate strategy*. New York, NY: Oxford University Press.

Creswell, J.W. (2007). *Educational research: Planning, conducting and evaluating quantitative and qualitative research* (3rd ed.). Upper Saddle River, NJ: Prentice Hall.

Creswell, J.W. (2009). *Research design: Quantitative, qualitative and mixed methods approaches* (3rd ed.). London: SAGE.

Creswell, J.W. (2013). *Research design: Qualitative, quantitative and mixed methods approaches* (4th ed.). Los Angeles, CA: SAGE.

Dyer, J., Kale, P. & Sigh, H. (2001). How to make strategic alliances work. *MIT Management Review, 44* (4), 37–43.

Elvers, D. & Song, C.H. (2014). R&D cooperation and firm performance – Evaluation of partnering strategies in the automotive industry. *Journal of Finance and Economics, 2*(5), 185–193.

Filippetti, A. & Savona, M. (2017). University–industry linkages and academic engagements: Individual behaviours and firms' barriers. Introduction to the special section. *The Journal of Technology Transfer, 42*(4), 719–729.

Gebrekidan, D.A. (2017). *The journey to and from international strategic alliances in emerging markets*. International Business in the Information Age – 43rd European International Business Academy Conference, 14–16 December 2017, Milan, Italy.

Guimon, J. (2013). *Promoting university–industry collaboration in developing countries.* Geneva, Switzerland: World Bank. doi:10.13140/RG.2.1.5176.8488
Gupta, S. (2005). *Research methodology and statistical technique.* New Delhi, India: Deep & Deep Publication.
Hagedoorn, J., Link, A.N. & Vonortas, N.S. (2000). Research partnership. *Research Policy, 29*, 567–586.
Iqbal, A.M. (2013). *Evaluation of research collaboration between university and industry* (Master's thesis, Faculty of Management and Human Resource Development, Universiti Teknologi Malaysia, Johor Bahru, Malaysia). Retrieved from https://core.ac.uk/download/pdf/42908459.pdf
Kamin, Y.B. (2012). *TAFE in Australia and community college in Malaysia compared: How are students prepared for the workplace in mechanical engineering (automotive)* (Doctoral thesis, Faculty of Education, La Trobe University, Melbourne, Australia).
Kapil, P. (2014). Bridging the industry–academia skill gap. A conceptual investigation with special emphasis on the management education in India. *Journal of Business and Management, 16*(3), 8–13.
Kauppila, O., Mursula, A., Harkonen, J. & Kujala, J. (2015). *Evaluating university– industry collaboration: The European Foundation of Quality Management excellence model-based evaluation of university–industry collaboration.* Industrial Engineering and Management. University of Oulu: Finland
Koschatzky, K. & Stahlecker, T. (2010). New forms of strategic research collaboration between firms and universities in the German research system. *International Journal of Technology, Transfer and Commercialisation, 9*(1–2). doi:10.1504/IJTTC.2010.029427
Lemos, D. & Cario, S.A. (2017). University–industry interaction in Santa Catarina: Evolutionary phases, forms of interaction, benefits and barriers. *Innovation and Management Review, 14*(1), 16–29.
Lendrum, T. (2003). *The strategic partnering handbook: The practitioners' guide to partnerships and alliance* (4th ed.). Sydney, Australia: McGraw-Hill.
Malaysian Employers Federation. (2004). *The MEF salary and fringe benefits survey for executives 2003.* Kuala Lumpur, Malaysia: Malaysian Employers Federation.
Malik, K. & Wickramasinghe, V. (2015). *Initiating university–industry collaborations in developing countries.* Paper presented at 5th Annual International Conference on Innovation & Entrepreneurship (IE 2015), Global Science & Technology Forum, Singapore. doi:10.5176/2251-2039_IE15.5
Muthoka, R.K. & Kilika, J. (2016). Towards a theoretical model for strategic alliance and partner selection among Small Medium Enterprises (SMEs): A research agenda. *Science Journal of Business and Management, 4*(1), 1–7.
Nevin, M. (2014). *The strategic alliance handbook: A practitioners guide to business-to-business collaborations* (1st ed.). Abingdon, UK: Routledge.
Patel, P. & D'Este, P. (2007). University–industry linkages in the UK: What are the factors underlying the variety of interaction with industry? *Research Policy, 36*, 1295–1313.
Pett, T.L. & Dibrell, C.C. (2001). The process model of global strategic alliances formation. *Business Process Management Journal, 7*(4), 349–364.
Philbin, P.S. (2008). Developing and managing university–industry research collaborations through a process methodology/industrial sector approach. *Journal of Research Administration, 11*(3), 51–68.
Rashidi, R.B.H. (2013). *Evaluation of collaboration between public training institutions and private industries and its importance in improving the quality of training delivery in TVET in Malaysia.* Bintulu, Malaysia: Advanced Technology Training Center (ADTEC). Retrieved from http://hdl.voced.edu.au/10707/256158.
Russo, M. & Cesarani, M. (2017). Strategic alliance success factors: A literature review on alliance lifecycle. *International Journal of Business Administration, 8*(3), 1–9.
Schanan, J.L. & Kelly, M.J. (2007). *Cooperative strategy.* New York, NY: Oxford University Press.
Shenkar, O. & Reuer, J.J. (2006). *Handbook of strategic alliances.* Thousand Oaks, CA: SAGE Publications.
Sidek Mohd Noah. (2002). Research Design: Philosophy, theory and practical. Serdang:Universiti Putra Malaysia Publisher.
Taratukhin V., Yury V., Becker J. (2016). Towards a Framework for Educational University-Industry Cooperation: Industry Perspective. ASEE 123rd Anual Conference and Exposition. America Society Engineering Education. New Orlean.
Thune, T. (2011). Success factors in higher education–industry collaboration: A case study of collaboration in the engineering field. *Tertiary Education and Management, 17*(1), 31–50.
Tjemkes, B., Vos, P. & Burger, K. (2012). *Strategic alliance management.* Abingdon, UK: Routledge.
Wahyuni, S. (2003). *Strategic alliance* development: *A study on alliances between competing firms* (Doctoral thesis, University of Groningen, The Netherlands).

TVET Towards Industrial Revolution 4.0– Hazirah Noh@Seth et al. (eds)
© 2020 Taylor & Francis Group, London, ISBN 978-0-367-24273-2

Effect of the just-in-time teaching technique on students' achievement and knowledge retention in elementary structural design in colleges of education in Nigeria

B.D. Beji
Department of Building Technology, Niger State College of Education, Minna, Nigeria

ABSTRACT: The study investigated the effects of the just-in-time teaching technique on students' achievement and knowledge retention in elementary structural design in colleges of education in Kwara and Niger States, Nigeria. Two research questions were raised and two hypotheses were formulated and tested at a 0.05 level of significance. A pretest–post-test non-equivalent control group design of a quasi-experimental study was used. The population for the study consisted of 128 Nigerian Certificate in Education (NCE) II technical students. A simple random sampling technique was used to allocate Kwara State College of Education (technical), Lafiagi to the control group and Niger State College of Education, Minna to the experimental group. The instrument used for data collection was the Elementary Structural Design Achievement and Retention Test (ESDART). The instrument was validated by two experts from the Federal University of Technology, Minna and one expert from the test development unit of the National Examination Council, Minna. The reliability coefficient and the internal consistency of the instrument were found to be 0.84 and 0.87, respectively, using the Pearson product moment correlation and Kuder–Richardson 20 formulas. The mean was used to answer the research questions while ANCOVA was employed to test the hypotheses. The findings of the study revealed that the just-in-time teaching technique was more effective in improving students' achievement and knowledge retention in elementary structural design than the conventional lecture method. It is therefore recommended that teachers teaching elementary structural design should adopt the just-in-time teaching technique to enhance students' achievement and knowledge retention.

1 INTRODUCTION

Elementary structural design is a compulsory two-credit unit course among other compulsory courses in technical education in Nigerian colleges of education. Elementary structural design is aimed at introducing Nigerian Certificate in Education (NCE) II technical students to the preliminary analysis of the structural design of building components and frames. According to the National Commission for Colleges of Education (NCCE) (2006), elementary structural design serves as an introductory course to statistics and the strength of materials, including the properties of materials used in residential and commercial building construction. Elementary structural design could be seen as a study of building static loads referencing concrete, steel, wood and pre-engineering wood products. Omeje (2013) revealed that the low performance of students in elementary structural design could be attributed to the analytical nature of the course and the persistent use of the conventional lecture method in colleges of education.

The conventional lecture method is a teacher-centered method of teaching that gives little or no attention to students' activities in the classroom. Adunfe (2005) defined a conventional lecture method as a teaching strategy where pre-packaged instructional content is delivered by the teacher to a large audience with minimal student–teacher interaction. The conventional lecture method of teaching has been persistently used for years in higher education institutions.

According to Jackson (2012), the conventional lecture method is now gaining poorer results when compared to more modern and revolutionary teaching methods. Furthermore, Ade (2000) articulated that the challenge posed by the conventional lecture method of teaching necessitates a shift to active teaching and learning techniques such as just-in-time teaching.

The Just-in Time Teaching (JiTT) technique is a teaching and learning technique that makes use of students' responses to web-based questions covering upcoming course material. According to Simkins and Maier (2010), the JiTT technique is a strategy that utilizes web-based technology to foster active learning. The JiTT technique uses feedback between classroom activities and assignments in preparation for the classroom session. In responding to these assignments, students usually have to read, watch or carry out activities and answer questions related to the assignments. Marrs (2010) revealed that the JiTT technique allows students to take control of their own learning process, gain motivation and enhance their performance. Martinez (2012) further noted that students come to class better prepared for the subject and teachers come to class better prepared for their students. The preparation of both the students and the teacher is an ingredient that leads to better student academic achievement.

Students' academic achievement means performance in a school subject as represented by scores or marks measured using an achievement test. According to Ogbu (2008), academic achievement refers to students' success in learning specified curriculum content. Tansel (2012) stated that academic achievement is defined as the learning outcomes of a student, which include the knowledge, skills and ideas acquired through the course of studies within and outside the classroom situation. According to Olatoye and Aderogba (2011), academic achievement is connected to the retention of knowledge.

Retention of knowledge is the ability of an individual to reproduce valuable knowledge after a period of time. According to Adamu (2016), retention of knowledge is the repeat performance by a learner of earlier acquired behavior, elicited after an interval of time. Retention simply refers to how much a person remembers after an interval of time without practice, and is the difference between what is initially learned and what is later forgotten (Chi, 2011). According to Martinez (2012), use of the JiTT technique has been shown to enhance students' academic achievement and knowledge retention in sciences and humanities. The JiTT technique may also enhance students' academic achievement and knowledge retention in elementary structural design.

1.1 *Statement of the Problem*

The low academic performance of NCE II technical students in elementary structural design was traced by Omeje (2013), who revealed that the low performance of students in elementary structural design could be attributed to the analytical nature of the course and the persistent use of the conventional lecture method in colleges of education. Oranu (2003) observed that most higher education institutions in Nigeria used conventional lecture methods in teaching. The shortcomings of the conventional lecture methods of teaching could be accountable for the low academic achievement and knowledge retention of students in elementary structural design. The question for this study is: what will be the effect of the just-in-time teaching technique on students' achievement and knowledge retention in elementary structural design in colleges of education?

1.2 *Aim and Objectives of the Study*

The aim of the study was to investigate the effects of the JiTT technique on students' achievement and knowledge retention in elementary structural design in colleges of education. Specifically, the objectives of the study were to determine:

1. The effect of the JiTT technique on students' achievement in elementary structural design;
2. The effect of the JiTT technique on students' knowledge retention in elementary structural design.

1.3 *Research Questions*

The following research questions guided the study:

1. What is the effect of the JiTT technique on students' achievement in elementary structural design?
2. What is the effect of the JiTT technique on students' knowledge retention in elementary structural design?

1.4 *Hypotheses*

The following null hypotheses were formulated and tested at a 0.05 level of significance:

H_{01}: There is no significant difference between the mean achievement scores of students taught elementary structural design using the JiTT technique and those taught with the conventional lecture method.

H_{02}: There is no significant difference between the mean knowledge retention scores of students taught elementary structural design using the JiTT technique and those taught with the conventional lecture method.

2 METHODOLOGY

A pretest–post-test non-equivalent control group design of a quasi-experimental study was adopted for the study. According to Nworgu (2006), random assignment of subjects in such a design is not required. The design is symbolically represented as follows:

Experimental O_1 X O_2 Y O_3

Control O_1 – O_2 Y O_3

where:

O_1: Represents the pretest (observation or measurement before treatment);

O_2: Represents the post-test (observation or measurement after treatment);

O_3: Represents the knowledge retention test (conducted two weeks after the post-test);

X: Represents the experimental treatment (JiTT technique); "–" indicates no treatment (that is, the use of a conventional lecture method);

Y: Indicates a delayed period of two weeks after the post-test.

The study was conducted in Kwara and Niger States, Nigeria. The population for this study comprised of 128 NCE II technical students in the 2017/2018 session. A simple random sampling technique was used to allocate Kwara State College of Education (technical), Lafiagi to the control group and Niger State College of Education, Minna to the experimental group. The instrument used for data collection was the Elementary Structural Design Achievements and Retention Test (ESDART). The instrument was developed by the researcher and consisted of 60 multiple-choice items with four response options based on the NCE technical syllabus for elementary structural design (TEB 214).

The researcher prepared two sets of lesson plans for teaching elementary structural design. Each set contained eight lesson plans that lasted for a period of eight weeks for a minimum of two hours per week. One set of the lesson plans was developed on the basis of JiTT technique while the second set was prepared according to the conventional lecture method.

The instrument was subjected to content validation by three experts that included two lecturers from the Department of Industrial and Technology Education, Federal University of Technology, Minna, Nigeria and one staff member of the Examination Development Department, National Examination Council (NECO). To ensure the reliability of the instrument, a trial test was conducted using 24 NCE III technical students in the Federal College of Education (technical), Bichi, Kano State, Nigeria. The split-half reliability technique was used on the instrument and the reliability coefficient and the internal consistency of the instrument were found to be 0.84 and 0.87, respectively, using the Pearson product moment correlation and Kuder–Richardson 20 formulas. Nevertheless, item analysis was carried out on the 60 items developed in the instrument to determine the difficulty and discrimination indices of each item in the test.

2.1 *Experimental Procedure*

The study involved three main stages, which included the administration of the pretest, post-test and retention test. The study was conducted for a period of ten weeks during which eight topics in elementary structural design were covered. The pretest was administered in the first week of the research exercise to all the students before the experimental and control groups were subjected to the treatments. The teaching process lasted for eight weeks after which the post-test was administered to all the students to determine their mean achievement. Two weeks after the post-test, a knowledge retention test was administered to all the students to determine their mean retention scores.

2.2 *Method of Data Analysis*

The data collected for this study were analyzed using Statistical Package for Social Sciences (SPSS) version 20.0 software. Descriptive and inferential statistics were used to analyze the data. The general Linear Model (univariate) function was used to perform the Analysis of Covariance (ANCOVA). The group with the higher mean value was taken to have performed better in the achievement or knowledge retention test in elementary structural design. With a significance of F less than 0.05, the null hypotheses were rejected.

3 RESULTS

Research Question 1

What is the effect of the JiTT technique on students' achievement in elementary structural design?

Table 1 shows that the experimental group had a mean achievement score of 14.30 in the pretest and a mean score of 52.23 in the post-test, with a pretest-to-post-test mean gain of 37.93. The control group had a mean achievement score of 14.27 in the pretest and a post-test mean of 37.33, with a pretest-to-post-test mean gain of 23.06.

Table 1. Mean of pretest and post-test achievement test scores of students taught elementary structural design using the just-in-time teaching technique and those taught by the conventional lecture method.

Group	N	Pretest mean	Post-test mean	Mean gain
Experimental	68	14.30	52.23	37.93
Control	60	14.27	37.33	23.06

Research Question 2

What is the effect of the JiTT technique on students' knowledge retention in elementary structural design?

Table 2 shows that the experimental group had a mean score of 14.30 in the pretest and a mean retention score of 48.54, with a pretest-to-retention test mean gain of 34.24. The control group had a mean score of 14.27 in the pretest and a mean retention score of 30.83, with a pretest-to-retention test mean gain of 16.56.

Hypothesis One

Table 2. Mean of pretest and retention test scores of students taught elementary structural design using the just-in-time teaching technique and those taught by the conventional lecture method.

Group	N	Pretest mean	Retention test mean	Mean gain
Experimental	68	14.30	48.54	34.24
Control	60	14.27	30.83	16.56

H_{01}: There is no significant difference between the mean achievement scores of students taught elementary structural design using the JiTT technique and those taught by the conventional lecture method.

Table 3 shows the F-calculated value for the effect of instructions on the cognitive achievement of students taught elementary structural design using the JiTT technique and those taught by the conventional lecture method. The F-calculated value for the groups is 113.429 with a significance for F of 0.00, which is less than 0.05. The results indicated that there was a statistically significant difference between the mean achievement scores of students taught elementary structural design using the JiTT technique and those taught by the conventional lecture method. Hence, the null hypothesis was rejected

Table 3. Summary of the Analysis of Covariance (ANCOVA) to test for the significant difference between the mean achievement scores of students taught elementary structural design using the JiTT technique and those taught by the conventional lecture method.

Source	Type III sum of squares	df	Mean square	F	Sig.
Corrected model	7,078.714[a]	2	3,539.357	5,876.549	0.000
Intercept	240.836	1	240.836	399.871	0.000
Pretest	0.283	1	0.283	0.470	0.494
Group	7,059.039	1	7,059.039	113.429	0.000
Error	75.286	125	0.602		
Total	269,242.000	128			
Corrected Total	7,154.000	127			

a R-squared = 0.989 (Adjusted R-squared = 0.989)
* Significance (F less than 0.05)

Hypothesis Two

H_{02}: There is no significant difference between the mean retention scores of students taught elementary structural design using the JiTT technique and those taught by the conventional lecture method.

Table 4 shows the F-calculated value for the effect of instructions on the knowledge retention of students taught elementary structural design using the JiTT technique and those taught by the conventional lecture method. The F-calculated value for the groups is 610.628 with a significance for F of 0.00, which is less than 0.05. The results indicated that there was a statistically significant difference between the mean retention scores of students taught elementary structural design using the JiTT technique and those taught by the conventional lecture method. Hence, the null hypothesis was rejected.

Table 4. Summary of the Analysis of Covariance (ANCOVA) to test for the significant difference between the mean retention scores of students taught elementary structural design using the JiTT technique and those taught by the conventional lecture method.

Source	Type III sum of squares	df	Mean square	F	Sig.
Corrected model	9,998.482[a]	2	4,999.241	3,048.172	0.000
Intercept	190.091	1	190.091	115.903	0.000
Pretest	0.191	1	0.191	0.117	0.733
Group	9,972.706	1	9,972.706	610.628	0.000
Error	205.010	125	1.640		
Total	217,491.000	128			
Corrected Total	10,203.492	127			

a R-squared = 0.980 (Adjusted R-squared = 0.980)
* Significance (F less than 0.05)

4 FINDINGS OF THE STUDY

1. The students taught elementary structural design using the JiTT technique obtained a higher mean achievement score than the students taught by the conventional lecture method.
2. The students taught elementary structural design using the JiTT technique obtained a higher mean retention score than the students taught by the conventional lecture method.
3. There was a significant difference in the mean achievement scores of students taught elementary structural design using the JiTT technique and those taught by the conventional lecture method.
4. There was a significant difference in the mean retention scores of students taught elementary structural design using the JiTT technique and those taught by the conventional lecture method.

5 DISCUSSION OF FINDINGS

The results presented on students' mean achievement scores in elementary structural design revealed that the students taught elementary structural design using the JiTT technique obtained a higher mean achievement score than the students taught by the conventional lecture method in the elementary structural design achievement test. This finding is in line with those of Gavrin (2010), who revealed that the JiTT technique increased student participation, preparation and achievement. Moreover, this finding was confirmed by Cookman (2010), who asserted that students enrolled in a course that successfully implements the JiTT technique will gain both problem-solving skills and conceptual understanding.

The summary of the Analysis of Covariance (ANCOVA) to test for the significant difference between the mean achievement scores of students taught elementary structural design using the JiTT technique and those taught by the conventional lecture method revealed a statistically significant difference. This finding conforms with that of Marrs and Novak (2004), who found a significant difference in the mean achievement scores of students in the experimental group and the control group in their study into the effect of the JiTT technique on student learning and student success in biology.

The results presented on students' mean retention scores in elementary structural design revealed that the students taught elementary structural design using the JiTT technique obtained a higher mean retention score in the Elementary Structural Design Achievement and Retention Test (ESDART) than the students taught by the conventional lecture method. This finding confirmed that of Martin (2016), who revealed that active web-based learning enhances students' retention of knowledge.

The summary of Analysis of Covariance (ANCOVA) to test for the significant difference between the mean retention scores of students taught elementary structural design using the JiTT technique and those taught by the conventional lecture method revealed that there was a significant difference. This finding conforms with that of Ozden (2015), who found a significant difference in the mean retention scores of students in the experimental group and the control group in their study.

6 CONCLUSION

The results obtained revealed that the academic achievement and knowledge retention of students were enhanced using the JiTT technique. Consequently, the JiTT technique has the potential to enhance learning via the internet by actively involving students in the pre-class activities and minimizing the teacher's involvement in the teaching process. Therefore, it is concluded that the JiTT technique has a positive effect on students' achievement and knowledge retention in elementary structural design in colleges of education.

7 RECOMMENDATIONS

Based on the findings, the following recommendations are made:

1. Teachers, especially those teaching elementary structural design, should adopt the JiTT technique to teach students at colleges of education to enhance students' achievement and knowledge retention in elementary structural design.
2. Teachers and students of elementary structural design should break away from the old method of teaching and ensure the instruction in colleges of education becomes active learning oriented as exemplified in the use of the JiTT technique.

REFERENCES

Adamu, S. (2016). Effects of multimedia instruction on students' achievement and retention in mathematics in Niger state. *Eurasia Journal of Mathematics, Sciences and Technology Education, 3*(1), 112–120.

Ade, I. (2000). *Principles and practice of teaching*. Nasarawa, Nigeria: Deegees Computers.

Adunfe, W. (2005). *The hearth of teaching*. Ibadan, Nigeria: Citadel Books.

Chi, M.T.H. (2011). *Learning from human tutoring cognitive science*. Retrieved from http://www.pitt.edu/Nchi/papers/image3.pdf

Cookman, C. (2010). Using just-in-time teaching to foster critical thinking in a humanities course. In S.P. Simkins & M.H. Maier (Eds.), *Just-in-time teaching* (pp. 163–178). Sterling, VA: Stylus Publishing.

Gavrin, A.D. (2010). *Using just-in-time teaching in the physical sciences*. London: Edward Arnold.

Jackson, S. (2012). *Three new teaching methods improve the education process*. Retrieved from http://gettingsmart.com/2012/09/categories/edtech

Marrs, K.A. (2010). Using just-in-time teaching in the biological sciences. In S.P. Simkins & M.H. Maier (Eds.), *Just-in-time teaching* (pp. 81–100). Sterling, VA: Stylus Publishing.

Marrs, K.A. & Novak, G. (2004). Just-in-time teaching in biology: Creating an active learner classroom using the internet. *Cell Biology Education, 2*(3), 49-61.

Martin, A. (2016). Assessing the effect of constructivist YouTube video instruction in the spatial information sciences on student engagement and learning outcomes. *Irish Journal of Academic Practice, 5*(1), 9–12.

Martinez, A. (2012). Using JiTT in a database course. *Proceedings of the 43rd ACM technical symposium on Computer Science Education*. Retrieved from http://www.researchgate.net/publication/254006871_Using_JITT_in_a_database_course

NCCE. (2006). *Minimum standards for Nigerian certificate in education teachers*. Abuja, Nigeria: National Commission for Colleges of Education.

Nworgu, B.G. (2006). *Educational research: basic issues and methodology* (2nd ed.). Enugu, Nigeria: University Trust Publishers.
Ogbu, J.U. (2008). *Minority status, oppositional culture, & schooling*. New York, NY: Routledge.
Olatoye, R.A. & Aderogba, A.A. (2011). Effect of co-operative and individualized teaching methods on senior secondary school students' achievement in organic chemistry. *The Pacific Journal of Science and Technology, 12*(2), 310–319.
Omeje, H.O. (2013). *Effects of two models of problem-based learning approaches on students' achievement, interest and retention in elementary structural design* (Doctoral dissertation, University of Nigeria, Nsukka, Nigeria).
Oranu, R.N. (2003). *Vocational and technical education in Nigeria*. Retrieved from http://www.ibec.unesco.org
Ozden, M. (2015). P effects of cognitive apprenticeship techniques learning on achievement and retention of knowledge in science course. *Eurasia Journal of Mathematics, Sciences and Technology Education, 3* (2), 157–161.
Simkins, S.P. & Maier, M.H. (2010). *Just-in-time teaching: Across the disciplines, across the academy*. Sterling, VA: Stylus Publishing.
Tansel, A. (2012). Determinants of school attainment of boys and girls in Turkey: Individual, household and community factors. *Economics of Education Review, 21*(2), 455–470.

TVET Towards Industrial Revolution 4.0– Hazirah Noh@Seth et al. (eds)
© 2020 Taylor & Francis Group, London, ISBN 978-0-367-24273-2

Constraints faced by technical teachers in the application of higher order thinking skills in the teaching process at vocational colleges

H.P. Kong & M.H. Yee
Department of Technical and Vocational Education, Parit Raja, Johor, Malaysia

ABSTRACT: Higher Order Thinking Skills (HOTS) play an important role in higher education learning. HOTS are a set of skills that use a concept that encourages individuals, especially students, to think outside the box. In other words, HOTS represent different individual preferences and strengths in learning and can be a stimulus for developing new ways of learning. The purpose of this research was to identify the problems faced by technical teachers in applying HOTS in the teaching process at vocational colleges. A total of four respondents from vocational colleges were randomly selected as samples. This is a qualitative research using only interviews to explore the perspective of teachers when applying HOTS in their teaching process, their perspective on the application of HOTS in vocational subjects and the constraints they faced in applying HOTS. The findings indicated that they faced several constraints in the application of HOTS in their teaching process at vocational colleges. It was found that it is not suitable to apply HOTS for vocational subjects, as the technical teachers lack knowledge of HOTS and also lack the training needed to be able to apply HOTS in vocational subjects.

1 INTRODUCTION

Along with the Technical and Vocational Education and Training (TVET) transformation, the TVET education system, such as vocational high schools, has been transformed into vocational colleges. In 2012, 15 vocational high schools were upgraded to vocational colleges. The curriculum implementation in vocational colleges is considered to be new to technical teachers. Understanding the implementation of the new curriculum is important. According to Azis (2013), the change in the education system in Malaysia is based on current needs and the level of development needed to move toward becoming a developed nation.

Hence, Higher Order Thinking Skills (HOTS) were introduced by the Ministry of Education (MOE, 2013) in 2013 in all areas of education, including vocational education, in order to reinforce the continuing critical and creative thinking skills among students. HOTS is now defined by the MOE as the ability to adapt knowledge, skills and values in reasoning and reflection in order to find solutions, solve problems, and use innovation and creativity to create something new. Thinking skills is one of the most important elements for developing intelligence, creativity and innovation among human capital in order to meet the challenges of the 21st century, so that the country can compete globally.

Teachers need to be the first to master these skills and they have to deepen their knowledge of HOTS before practicing or applying it in the learning process. Thus, a supply of technical teachers with the ability to apply HOTS ahead of the learning process is important, especially the use of problem-solving techniques in practical training (Zulkifli, 2016). So, technical teachers need to create and revise existing information with various references and look for information regarding HOTS to ensure optimum control in the field before practicing it.

This is because the application of HOTS in the field of education is crucial in order to improve students' thinking skills before they enter the field of work.

2 PROBLEM BACKGROUND

TVET is based on competence; therefore, the skills implemented are considered to be more difficult to deliver because they need to be taught and assessed more specifically than other academic subjects (Sukri, 2013). However, Azman (2016) highlighted that those students who are admitted to vocational colleges are mostly those students who are weak in academic subjects, because many regard the vocational field as being suitable only for those who are academically weak. Because of this, various problems are faced by the teachers and students during the teaching and learning sessions at vocational colleges. Among the problems raised were the students' weak mastery of the theory, insufficient facilities and equipment, the students skipping classes, and so on (Rahman et al., 2015). These problems contribute to the constraints found in the application of HOTS in vocational colleges in order to produce skilled and knowledgeable human capital as listed in the Ninth Malaysia Plan, which aims to enhance the capacity of national knowledge and innovation (Asnul Dahar et al., 2013). The recent Eleventh Malaysia Plan for 'Accelerating Human Capital Development for an Advanced Nation' has listed industry-led TVET.

The lack of knowledge among teachers regarding HOTS (Kassim & Zakaria, 2015) is also one of the main causes for the weak mastery and application of thinking skills among students. Abdullah et al. (2015) found that most teachers understand the theoretical cognitive level of Bloom's taxonomy, but they cannot understand the different functions of each level in the taxonomy. Because of their lack of knowledge of HOTS, teachers are worried about their ability to apply HOTS in class. Based on teacher interviews in the study conducted by Sulaiman et al. (2017), they are more concerned about the ability of teachers to implement HOTS, as many teachers are still not capable of applying HOTS in class, and they also need more exposure and training to master the application of HOTS in lessons. This finding is supported by the requirement study report by Kestrel Education Consultants (UK) and 21st Century School (USA) (2011), which shows that the level of HOTS application among teachers and students in Malaysia is very low (Hasan & Mahamod, 2016).

Therefore, the aim of this study is to identify the constraints faced by teachers in the application of HOTS in the teaching process at vocational colleges.

3 RESEARCH OBJECTIVES

The objectives of this study are to identify:

1. The perspective of teachers in applying HOTS in the teaching process.
2. The perspective of technical teachers in applying HOTS in vocational subjects.
3. The constraints faced in the application of HOTS in the teaching process at vocational colleges.

4 METHODOLOGY

This research used a qualitative approach. The method used to obtain the results for this research involved conducting interviews at vocational colleges. The interviews involved four experts (E1, E2, E3 and E4) in their fields at vocational college. They were familiar with the field and had experience in the process of teaching and learning in vocational colleges. The interviews were conducted to collect information about the constraints faced in the application of HOTS in the teaching process at vocational colleges. Each interview session lasted for about 30 minutes. The feedback was transcribed and categorized, and finally themes were formed based on the categories (Sulaiman et al., 2017).

5 FINDINGS

The findings of this research were analyzed and discussed on the basis of the research objectives.

5.1 *The Perspective of Teachers in Applying HOTS in the Teaching Process*

Most of the teachers had three major perspectives with regard to HOTS. First, they had a positive view of the implementation of HOTS in the teaching and learning process because 'It is good for students' (E1) and 'Students can learn how to use their thinking skills wisely' (E2). Nowadays, life is full of challenges, where 'Students need to train themselves to think critically and more creatively' (E3) and 'Students need to know how to generate ideas based on their own knowledge' (E4). However, they were confused and worried about the ability of technical teachers to implement HOTS effectively, and in vocational subjects, because 'Teachers did not receive any training on HOTS' (E2), 'Teachers only have a very basic knowledge about HOTS' (E4), 'Teachers do not have enough time to learn HOTS' (E1) and 'HOTS is more suitable to be implemented in theoretical subjects' (E3).

5.2 *The Perspective of Technical Teachers in Applying HOTS in Vocational Subjects*

Vocational subjects focus more on practical lessons: 'Normally the vocational subjects apply hands-on teaching because they have a lot of complicated procedures' (E3); 'Vocational Colleges are more focused in marketability, which means students can work with their skills, therefore thinking skills is not suitable' (E1). During practical lessons, students need to spend a lot of their time on finishing a lab report or project: '... Maybe teachers can use HOTS to ask students based on real-life situation because this will help students apply HOTS' (E4); 'Thinking tools like I-think and mind mapping are good for vocational students who are weak in analyzing the concepts and facts' (E2).

5.3 *The Constraints Faced in the Application of HOTS in the Teaching Process at Vocational Colleges*

There are three main constraints in the application of HOTS in the teaching process at vocational colleges. The first constraint is that 'Technical teachers do not receive any exposure and training on HOTS application in lessons' (E1 & E2). The second constraint is that there are 'Different level of students in learning process' (E4) and 'HOTS is suitable and more effective for students who are excellent in study' (E2). The last constraint in the application of HOTS in the teaching process is that 'The responsibility of technical teachers are heavy and they have no time to learn and apply HOTS in vocational subjects because teaching vocational subjects is not an easy job' (E3).

6 DISCUSSION AND CONCLUSION

Based on the aspirations of the Malaysian Education Blueprint 2013–2025, teachers generally understand that HOTS need to be infused through lessons in order for students to achieve better results. Thus, these findings are in line with Sulaiman et al. (2017) and Nessel and Graham (2007), who tried to infuse HOTS into both academic and real-life situations.

The respondents understood the benefits of applying HOTS in vocational subjects. Based on Harland (2002) and Gordon et al. (2001), several effective strategies, such as questioning and concept maps, are the main strategies used to apply student-centered and problem-based approaches in teaching. These approaches are able to create an avenue for students to think and challenge themselves in the learning process (Sulaiman et al., 2017; Bissell & Lemons, 2006).

In applying HOTS in teaching, teachers can, in fact, use quantitative and qualitative assessments to improve students' thinking and learning processes (Noor, 2008). Some examples of quantitative assessments are the use of a multiple-choice quiz, semi-structured questions, and occasional short and long essays to assess student learning; meanwhile, qualitative assessment methods include observation, interviews and discussions with students in order to assess their learning processes. The qualitative form of assessment seems to be the most authentic type of assessment (Sener et al., 2015). Therefore, the application of HOTS in the teaching process is essentially suitable for vocational subjects.

Therefore, in order to ensure the smooth application of HOTS in the teaching process in vocational subjects, the constraints highlighted need to be addressed. In order to tackle the first constraint, technical teachers should be exposed to HOTS and trained on how to apply them in the teaching process. According to Sharuji and Nordin (2017), most teachers have no guidelines on HOTS teaching methods. This problem consequently contributes to other limitations faced by the teachers in applying HOTS teaching methods and strategies during class. Insufficient guidelines, books and modules on the application of HOTS in the teaching process will cause teachers to be anxious and hesitant in applying HOTS in the classroom, even though they have attended courses on HOTS (Rajendran, 2001).

According to Lee (2010), HOTS modules have many advantages for users. They are able to trigger someone to continue learning according to their own levels and abilities, choose the right way of learning, and identify their own strengths and weaknesses. Therefore, printed modules are more flexible for technical teachers who need to attend many training courses and programs. With this effective learning module, the last constraint can be reduced.

Consequently, teachers would be more able to improvise various settings and strategies to suit students' needs (Saido et al., 2015). Making an effort to use teaching strategies, such as group work, to balance the level of learning of the students will encourage the students to share and help each other, which will help those members who struggle with the learning process if the teaching process is too complicated or too fast. With the aims of improving the application of HOTS to be materialized, continuous and serious monitoring and the improvement of HOTS training must be undertaken.

ACKNOWLEDGMENTS

The authors would like to thank the Ministry of Higher Education, Malaysia for supporting this research under the Teacher Research Grant (TRG) No. VOT: V030. In addition, the authors also wish to thank the Faculty of Civil and Environmental Engineering (FKAAS), Universiti Tun Hussein Onn Malaysia (UTHM), which has given its full co-operation to ensure the success of this study.

REFERENCES

Abdullah, A.H., Aris, B., Saud, M. S., Boon, Y. & Awang Ali, S. A. (2015). *The implementation of high order thinking skills (HOTS): Issues and challenges in the aspects of curriculum, pedagogy and assessor.* In Prosiding Seminar Kebangsaan Majlis Dekan-Dekan Pendidikan Universiti Awam 2015, 14–15 September 2015, UTHM Johor.

Asnul Dahar, M., Ruhizan, M.Y., Kamalularifin, S. & Muhammad Khair, N. (2013). *The development of workshop strategy in education of technical dan vocational.* Paper presented at the 2nd International Seminar on Quality and Affordable Education (ISQAE 2013), KSL Hotel & Resort, Johor Bahru, Johor, Malaysia.

Azis, N.W. (2013). *High order thinking skills for Form 4 students in promblem solving.* Bachelor of Education diploma (Matematik). Faculty of Education. Universiti Teknologi Malaysia, Johor, Malaysia.

Azman, S.A. (2016). *Transformation of vocational high school to vocational colleges.* Sarjana Pentadbiran Pendidikan. Universiti Kebangsaan Malaysia, Selangor, Malaysia.

Bissell, A.N. & Lemons, P.P. (2006). *A new method for assessing critical thinking in the classroom.* BioScience, 56(1),66–72.

Gordon, P.R., Rogers, A.M., Comfort, M., Gavula, N. & McGee, B.P. (2001). *A taste of problem-based learning increases achievement of urban minority middle-school students.* Educational Horizons, *79*, 171–175.

Harland, T. (2002). *Zoology students' experiences of collaborative enquiry in problem-based learning.* Teaching in Higher Education, 7(1),3–15.

Hasan, N.H. & Mahamod, Z. (2016). *The perceptions of high school malay languange on high order thinking skills.* Jurnal Pendidikan Bahasa Melayu, 6(2),78–90.

Kassim, N. & Zakaria, E. (2015). *Intergrasion of high order thinking skills in process leaning dan teaching in subject mathematics: Teachers needs analysis*. Jurnal Pendidikan Malaysia, 30(2),15–32.
Lee, M.C. (2010). *Self teaching module for the topic concept electrolysis of form 4 KBSM based on master's learning strategies*. Johor, Malaysia: Universiti Teknologi Malaysia.
MOE. (2013). *Malaysia Education Blueprint 2013–2025 (Preschoool to Post Secondary Education)*. Putrajaya, Malaysia: Kementerian Pendidikan Malaysia [Malaysian Ministry of Education].
Nessel, D.D. & Graham, J.M. (2007). *Thinking strategies for student achievement, improving learning across the curriculum, K-12* (2nd ed.). Thousand Oaks, CA: Sage Publishing.
Noor, A.M. (2008). *Teaching thinking skills: Redesigning classroom practices*. Brunei: Universiti Brunei Darussalam.
Rahman, K.A., Saud, M. S., Kamin, Y. & Samah, N. A. (2015). *Problems in teaching and learning for electric technology courses at vocational colleges*. Jabatan Pendidikan Teknikal dan Kejuruteraan. Johor, Malaysia: Universiti Teknilogi Malaysia.
Rajendran, N.S. (2001). *Teaching high order thinking skills: The teacher's preparation to handle the process of teaching and learning*. Seminar/Pameran Projek KBKK: Warisan-Pendidikan-Wawasan. Kementerian Pendidikan Malaysia: Pusat Perkembangan Kurikulum.
Saido, G.M., Siraj, S., Nordin, A.B. & Al Amedy, O.S. (2015). *Higher order thinking skills among secondary students in science learning*. The Malaysian Online Journal of Educational Science, 3(3),13–20.
Sener, N., Turk, C. & Tas, E. (2015). *Improving science attitude and creative thinking through science education project: A design, implementation and assessment*. Journal of Education and Training Studies, 3(4),57–67.
Sharuji, W.N.S. & Nordin, N.M. (2017). *The teacher's preparation in application high order thinking skills (HOTS)*. Simposium Pendidikan diPeribadikan: Perspektif Risalah An-Nur (SPRiN2017). Faculty of Education, Universiti Kebangsaan Malaysia.
Sukri, N. (2013). *The level of readines of teachers towards the implemetation of compentence – based learning in vokasional colleges*. Bachelor of Education Teknik dan Vokasional, Faculty of Education, Universiti Teknologi Malaysia, Johor, Malaysia).
Sulaiman, T., Muniyan, V., Madhvan, D., Hasan, R., & Rahim, S.S.A. (2017). *Implementation of higher order thinking skills in teaching of science: A case study in Malaysia*. International Research Journal of Education and Sciences (IRJES), 1(1),1–3.
Zulkifli, Z. (2016). *Effectiveness of problem solving strategy in the improvement of high order thinking skills and solve problems among students*. Bachelor's thesis. Universiti Tun Hussein Onn Malaysia, Johor, Malaysia.

TVET Towards Industrial Revolution 4.0– Hazirah Noh@Seth et al. (eds)
© 2020 Taylor & Francis Group, London, ISBN 978-0-367-24273-2

Improving writing skills of eleventh-grade students by writing recount text through a field learning experience strategy

E. Mulyadi, A. Naniwarsih & S. Wulandari
Department of English Education, University of Muhammadiyah Palu, Indonesia

ABSTRACT: It is typically found that there are still many English teachers who cannot implement a proper strategy for teaching writing. The most crucial problem is that teachers lack an effective strategy to teach English writing skills in teaching the recounting of an event or experience. This research is focused on analyzing: (1) how to improve the teacher's teaching performance; (2) how to improve students' learning motivation and participation in the teaching–learning process; (3) how to improve students' achievement in English writing skills, particularly in writing recount text through a field learning experience strategy. The purpose of the research is to find an effective strategy for teaching English writing. The research employed a classroom action research method in a cyclic design that was implemented in spiral mode covering planning, implementation, observation and reflection. The research instruments were observation checklists, field notes, documentation, and testing; data was collected through cycles 1 and 2, each cycle involving three meetings. The subject of research was 24 eleventh-grade students at SMA Negeri 7 Palu. The data were analyzed both qualitatively and quantitatively. The findings of the research revealed that: (1) the teacher's teaching performance showed improvement from cycle 1 to cycle 2; (2) the students' learning motivation and classroom participation increased significantly; (3) the students' average score from cycle 1 was 63.7, and their classroom achievement was 45.8%; from cycle 2, their average score was 72.91 and their classroom achievement was 91.66%. The significance of the research is the increase in students' achievement and motivation in learning English writing skills.

1 INTRODUCTION

The skill of writing English is one of the language skills for which more attention should be paid to its teaching. Students can communicate with others or they can express their ideas and personal opinions through writing rather than speaking. According to Ehte Olshtain, in Celce-Murcia (2001, p. 207), writing as a communicative activity needs to be encouraged and nurtured during the language learner's course of study. This opinion is strengthened by Raimes (1983, p. 3), who gave the reasons for teaching writing as: "We frequently have to communicate with each other in writing", and "Writing reinforces grammatical structures, idioms, and vocabulary"; teaching writing is "a unique way to reinforce learning".

Unfortunately, it is typically found that there are still many English teachers who cannot implement a proper teaching method for writing. As a result, classes run monotonously and uninterestingly, which keeps students away from lessons. This is one of the problems found in the eleventh-grade students of SMA 7 Palu: the teacher lacks a writing teaching strategy, the students did not enjoy English, had low motivation to study, could not participate well in classroom learning activities and, of course, they attain low achievements in English writing skills.

Moreover, in reality, most students have difficulties in writing lessons. Based on a preliminary observation in the secondary high school, SMA 7 Palu class XI, we found that most students had problems when the teacher instructed them to write about their personal experience on holiday. They were confused as to what and how to write, because unclear instructions were delivered by the teacher in the teaching and learning process and the teacher implemented an inappropriate strategy during the lesson.

The data gained from this preliminary test showed that most students were categorized as having low achievement in writing lessons. Thus, 25% of the total of 24 students were categorized as having high achievement, 29.2% as middle achievement, and 45.8% as low achievement. It is not only students' achievement that is categorized as low, but also their classroom learning participation and motivation.

To overcome the students' problems in learning writing skills, we offered a field learning experience strategy to teach writing skills. This is a teaching strategy that involves the students in individual experiences gained from the field that enable them to construct a piece of writing. Another term to address such a strategy is field notes. Orion & Hofstein (1994), argued that the most important role of field trips in the learning process is the "direct experience with concrete phenomena and materials". The very nature of the field learning experience requires students to be active rather than merely learning in a passive mode in a traditional classroom.

In line with Orion & Hofstein (1994) and Bukian (2004) explains that, nowadays, the classroom teaching–learning process has become a burdensome activity, while outside the classroom it is a gratifying one. This does not mean that the use of the classroom for learning activities should be limited. It's one of the teaching strategies to introduce students to the real world, the field, for learning experiences. The core idea can be gleaned from the findings above; that classroom activities are an absolute must, but field-trip learning activities are one of the best possible approaches to take when classroom activities present boredom to students who study for a long time.

On the other hand, Finchum (2013) found that a well-organized and well-planned field trip can be educational, yet still be fun. While many school systems limit the number or even exclude field trips altogether, field trips, when possible, can have a deep and lasting educational value. Life-long memories can be created and a desire for learning can be sparked. Often it is the field trip that a student is likely to remember years later, more so than most activities and lessons completed in the classroom setting.

The above findings show that a field learning experience strategy encourages students to stay close to real-life fact learning where they can see things around them to improve their thinking skills and imagination, and leads them to put on paper what they see, feel, and think. Moreover, such direct-experience learning will empower the students with life-long memories concerning what they experienced.

A field learning experience strategy actively involves the student in learning in the real world, by visiting a place or object outside school to learn about or investigate it; for example, visiting the beach, the traditional market, supermarket, or park, which would make a big contribution to increasing students' motivation in their learning process. Moreover, the involvement and experience of the students in learning in the real world would enhance their ideas for writing because they learn in a more concrete and relaxed situation, through their own experience, and they will always be curious and enjoy learning about the presented topic. Besides, the field learning experience is one of the strategies available to the teacher to overcome the problem of teaching writing skills.

In order to direct our research to address the teacher's and students' problems in the teaching and learning of English writing skills, we phrased our research questions as follows: How to improve the teacher's teaching performance to teach English writing skills, particularly in writing a recount text through a field learning experience strategy? How to improve students' learning motivation and classroom participation through a field learning experience strategy? How to improve students' achievement in English writing skills, particularly in writing a recount text through a field learning experience strategy?

The focused content of teaching writing skills to senior high school students is to teach them about kinds of text – genre and language function. To teach kinds of text to students is not only accustoming them to reading fluently and accurately, but should also be training them how to construct a simple text. One of the texts that the curriculum suggests students study at the eleventh grade is recount text. Fauziati (2010, p. 53) explained that "Genre-based teaching is concerned with what learners do when they write. It allows [the] writing teacher to identify the kinds of texts that students will have to write in their target context (occupational, academic, or social) and to organize their courses to meet these needs". Genre-based text teaching provides more chances to students to explore different kinds of text, understanding its social function, generic structure and language features.

Meanwhile, recount text is assumed to be a simple text that the student can easily construct from their individual experience. Moreover, recount text is also considered a more appropriate kind of text to teach writing skills to students; because the students have different kinds of experience, so they can produce any kind of information they experienced in the field and place this in their writing content. In other words, the student will gain a variety of ideas for writing because they have a variety of experiences as well. In accordance with earlier statements, Syahmadi (2013, p. 79) stated that the English learning scope for eleventh-grade students is interpersonal discourse, transactional and functional, logic and coherent rhetoric; that is, the communication and the development of academic potential in five varieties of functional discourse: recount, narrative, procedure, description and report.

According to Anderson and Anderson (1997, p. 48), "Recount text is a piece of text that retells past events, usually in [the] order in which they happened." The purpose of a recount is to describe to an audience what occurred and when. In response to the above statement, it is easier for students to construct their ideas for writing because they write about what they experienced. It will enable them to advance the content of their writing from what they see, feel and think about while in the field.

To teach students writing skills means to enhance their ideas and thinking, to prompt them to share ideas, to encourage them to think critically, and to accustom them to the idea of delivery through writing. In teaching writing skills, the students are not just tasked with expressing ideas on paper but it simultaneously teaches them about kinds of text, grammatical aspects and mastery of vocabulary.

One of the methods used in teaching is the inquiry method. According to the dictionary, inquiry is questioning or investigation. According to Victor and Kellough (2004), in Jacobsen et al. (2009, p. 243), "Inquiry is a process [that] answers questions and solves problems based on knowledge and observations." In this model, students are given concrete materials and questions. In order to enable the students to answer the questions, they can work individually or in small groups to explore, observe, and discover answers. The teachers can then expand upon the discoveries by assigning the students to do further exploration in the field.

To teach writing skills, Brown (2001, p. 335) stated that written product is completed after the process of thinking, planning, drafting, and revising, and also demands efforts and specialized skills in generating ideas, organizing them coherently, making use of discourse markers and rhetorical conventions, putting all of these into one, revising the content for a clearer meaning, and editing for accurate grammar into a final product. In these terms, it can be said that one of the teacher's responsibilities in teaching English is to teach students writing skills in an appropriate procedure that will, in turn, accustom them to think critically before writing.

2 RESEARCH METHOD

The design of this research is Classroom Action Research (CAR). It is focused on eleventh-grade students using a qualitative and quantitative approach. This classroom action research was designed to develop students' writing skills in recount text through a field learning experience strategy.

The research was conducted collaboratively between the researchers and an English teacher. Cohen and Manion (in Nunan, 1989, p. 12) argue that, "Action research is closely related to the context in which it takes place, involves the collaboration of researcher as well as teachers, and is self-evaluative." Collaborative action research involves at least two persons as the main actors of the study action and this research team works together to solve the problem in a single classroom research activity. The researchers acted as collaborators and the English teacher in the school presented the teaching.

The research was conducted at SMA Negeri 7 Palu. This school is under the supervision of *Pemerintah Kota Palu*. The school has six parallel classes for each grade. The students were randomly distributed in each grade, to avoid unexpected competition in the classroom. Unexpected competition here means that high-achievement students were gathered into one class and so too were the low-achievement students. In order to create student interactions, the high- and low-achievement students were equally distributed in the same class.

The subjects of the research were second-year students of the even semester in the academic year 2017/2018. Because there were six parallel classes of second-year students, we took grade-XI IPA 3 students as the subjects of our research. Because the number of students representing the population of this research was small, just 24, they were all included in the research sample.

To do this research, we and the collaborating teacher began the research with a short discussion to determine the criteria for success. The results of this discussion determined that the criterion for success was that students should achieve an English writing skill score of 65. The score is aligned with the minimum criteria for achievement in English determined by the institution.

The procedure of conducting classroom action research covers four main steps; planning, implementing, observing, and reflecting. Planning is done by the researcher and the collaborating teacher to prepare the research instruments, arrange the lesson plans, prepare teaching materials, and design the evaluation form. Before implementing the planning, the researcher and the collaborating teacher set the scenario for the teaching and learning activities.

The implementation phase is the phase in which the researcher implemented the teaching and learning activities as prepared from the lesson plans, while the collaborating teacher conducted the observation. In the implementation, the researcher should follow the teaching procedure as it was arranged in the planning (see Section 3.1). The teaching–learning procedure consists of three phases; pre-activity, while-activity, and post-activity. The three phases should be completed in one session to measure the effectiveness of the implemented strategy and to identify the students' engagement in the learning process.

The observation phase is the phase in which the collaborating teacher observes the teaching–learning process conducted by the researcher. At the same time as teaching, the researcher conducted observation with regard to the students' learning activities and their involvement in the classroom tasks, which was categorized into two: students' classroom participation and students' learning motivation. The reflection phase is the phase for the researcher and the collaborating teacher to reflect on the planning, the previous teaching–learning process, and the observation activities.

In the reflection phase, the researcher and the collaborating teacher identified and analyzed the planning they did previously to identify anything that might have been omitted. If they found any teaching steps missing from the first cycle, they would complete them in the next one. The researcher and the collaborating teacher also analyzed the implementation of the strategy. If they found there were mistakes in implementing the strategy, they corrected them in the next cycle. In short, reflection is the place to review what has been done in the previous session or cycle and for better planning and implementation of the next session/cycle.

The cyclical procedure in classroom action research means that the research is conducted in cycles. The first cycle consisted of three meetings. The first meeting was oriented to introduce the teaching strategy, to accustom the students to write mechanically, and to conduct a small classroom writing task. The second meeting involved the field learning experience activities. The students were directed to visit previously determined places. In the field, the students identified any particular objects and they captured them in the form of field notes. Then, the last meeting was directed to evaluation. In this evaluation phase, the students were asked to construct a written draft involving recount text based on their individual experience of the previous meeting.

The second cycle also consisted of three meetings, as per the first cycle. There were some changes made by the researchers and the collaborating teacher before implementing the teaching–learning process in the second cycle. The changes were made according to the result of analysis in the reflection phase of the first cycle. The main change was focused on the implementation of teaching steps where, in the first cycle, the researcher omitted some. Besides the change in the teaching steps, the researchers and the collaborating teacher focused on the observations related to the students' learning motivation and classroom participation. They did a more detailed observation of the students' classroom participation and learning motivation in order to establish a more precise analysis of the change in these two aspects.

2.1 *Scheme of Classroom Action Research*

The researcher and the collaborating teacher were responsible for preparing teaching material and media. They prepared teaching material based on the students' abilities and media based on the skills focused upon in the research.

The criterion for success that the researcher and the collaborating teacher prepared were that the students should achieve the minimum criterion of achievement in writing skills, which was 65, based on that determined by the school for the subject of English. The class achievement should be determined from 75% of the total number of the students.

To implement the teaching of writing skills strategy well, the researcher and the collaborating teacher implemented a three-phase technique in the teaching–learning process, consisting of pre-activity, while-activity, and post-activity. Pre-activity was conducted by directing the students to the topic they are going to study. While-activity was the place where the researcher instructed the students on how to do the task and provided them with examples of class activities. Post-activity was the place for the researcher and the students to reflect on previous classroom activity and evaluate students' progress in writing skills.

The researcher and the collaborating teacher shared responsibility with each other in terms of observation activities. They used an observation checklist in performing their observation. The researcher observed students' participation, while the collaborating teacher observed the way the researcher implemented the teaching strategy and the students' responses and participation in the teaching–learning process.

The researchers and the collaborating teacher used four kinds of instruments in collecting data:

a. Field notes

These were used to record the teacher's and students' activities in the teaching and learning process during the research.

b. Observation checklist

For teachers, the steps in conducting the teaching and learning process;

For students', their English learning motivations and classroom participation.

c. Camera

For capturing classroom appearance and interaction.

d. Test

To gain information about the students' achievement in English writing skills;

The result of evaluation of the teaching of writing skills was determined via the following formula for determining students' final achievement, taken from the curriculum:

$$Score = \frac{Acheivement\ score}{Maximum\ score} \times 100$$

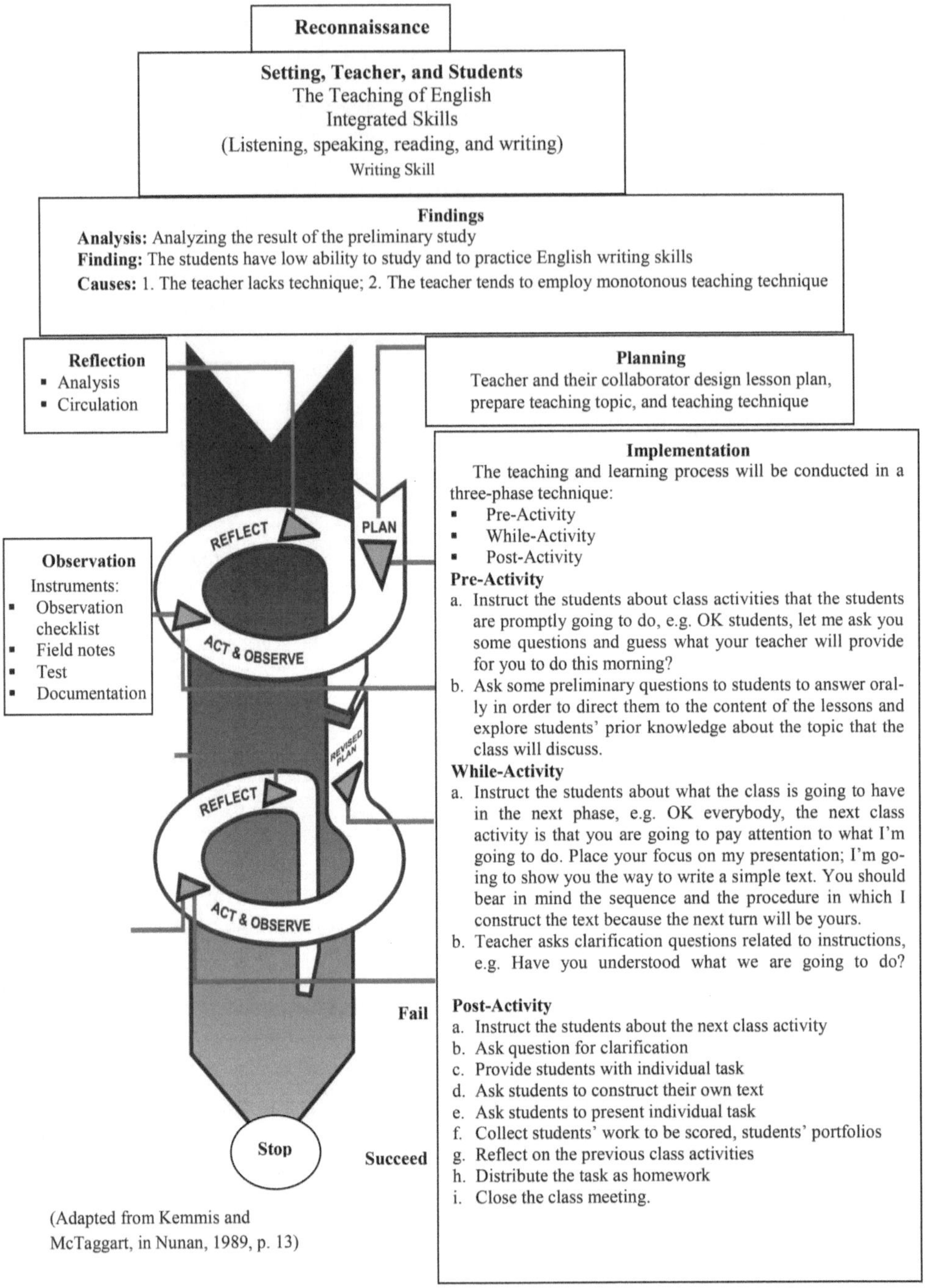

Figure 1. Scheme of classroom action research.

3 RESEARCH FINDINGS AND DISCUSSION

3.1 *Findings*

One of the main characteristics of classroom action research is not sole emphasis on the result of a treatment given in the classroom, but also consideration of how a treatment is given. This is one of the reasons for the research to present three kinds of data in this research: 1) teacher's teaching preparation and performance; 2) students' participation and motivation during the teaching and learning process; 3) the result of evaluation of students' English writing skills improvement.

3.1.1 *Planning*

Table 1 shows that in cycle 2 the researcher and the collaborating teacher prepared the same research instruments as in cycle 1: there was no change in the teaching and research preparation during the research.

3.1.2 *Implementation*

To implement the research, the researcher conducted the teaching steps shown in Table 2 in both cycles 1 and 2.

The teacher was required to conduct identical teaching steps in each session of cycles 1 and 2. To make the research run smoothly, (the number of steps were taken during the teaching–learning process. It is intended to find the effectiveness of the teaching procedure to encourage the students to study.

To conduct the teaching–learning process, the researchers conducted three sessions in each cycle. Each session consisted of opening the class, pre-activity, while-activity, and post-activity. As shown in Table 2, class-opening covered four activities, pre-activity involved four main activities, while-activity had 14 main activities, and post-activity had six main activities. While-activity is where the teacher introduced teaching materials, explained the kind of text (recount text; the characteristic of the text, social function of the text, generic structure, and language feature of the text), and the steps in writing. In the second session of each cycle, the students were divided into three groups and were directed to observe in the field. Each group had a different place/object to observe. To conduct the observation, the students were under the guidance of the collaborating teacher and a volunteer teacher.

Table 1. Teacher's teaching preparation.

No.	Prepared components	
	Cycle 1	Cycle 2
1.	Determine the criteria of success	Determine the criteria of success
2.	Lesson plan	Lesson plan
3.	Teaching materials/media	Teaching materials/media
4.	Field notes sheet	Field notes sheet
5.	Observation checklist/sheet	Observation checklist/sheet
6.	Camera	Camera

3.1.3 *Observation*

Observation was conducted by the researcher and the collaborating teacher. It was focused on teacher's teaching performance, students' learning motivation and participation, and the results of evaluation.

Table 2. The teaching steps.

No	Teaching steps	Teaching activities	Conducted	
			Yes	No
1	**Opening the class**	• Greeting the students		
		• Responding to students' greeting		
		• Asking students' situations		
		• Checking students' presence		
2	**Pre-Activity**	• Instructing the students as to what the class is going to do		
		• Asking students questions about the topic that is going to be learned to explore students' prior knowledge of it		
		• Providing students with the chance to ask clarification questions		
		• Responding to students' questions related to the planned topic		
3	**While-Activity**	• Giving instruction to students as to how to do the task		
		• Giving students the chance to ask clarification questions		
		• Giving brief explanation about the topic to students		
		• Providing students with the chance to clarify the teacher's explanation		
		• Continuing explanation		
		• Giving instruction about the kind of task that is going to be completed		
		• Giving instruction on how to complete the task		
		• Providing students with the chance to ask clarification questions		
		• Responding to students' questions		
		• Dividing students into groups or pairs		
		• Directing the students to the field		
		• Controlling the students' activities		
		• Asking the students to complete the classroom tasks/evaluation		
4	**Post-Activity**	• Giving instruction about the kind of task that is going to be completed		
		• Giving instruction on how to complete the task		
		• Providing students with the chance to ask clarification questions		
		• Responding to students' questions		
		• Providing feedback to students		
		• Administering homework		

3.1.3.1 TEACHER'S TEACHING PERFORMANCE

Table 3. Teacher's teaching performance.

No.	Cycle 1	Cycle 2
1.	**Opening the Class** All four activities in class-opening were conducted by the teacher (100%)	**Opening the Class** All four activities in class-opening were conducted by the teacher (100%)
2.	**Pre-Activity** The researcher conducted only two of four main activities suggested in pre-activity (50%)	**Pre-Activity** The researcher conducted all four main activities suggested in pre-activity (100%)
3.	**While-Activity** The researcher conducted eight of 13 main activities suggested in while-activity (53.3%)	**While-Activity** The researcher conducted all 13 main activities suggested in while-activity (100%)
4.	**Post-Activity** The researcher conducted two of six main activities suggested in post-activity (33.3%)	**Post-Activity** The researcher conducted all six main activities suggested in post-activity (100%)

3.1.3.2 STUDENTS' LEARNING MOTIVATION AND PARTICIPATION

Table 4. Students' learning motivation and participation.

No.	Observed aspects	Description	Cycle 1 (%)	Cycle 2 (%)
1	Classroom participation	a. The students pay serious attention to the teacher's explanation	45	90
		b. The students frequently ask clarification questions	25	70
		c. The students enjoy doing classroom tasks collaboratively	40	85
		d. The students readily provide help to classmates	15	75
2	Learning motivation	a. The students come to the classroom punctually	65	95
		b. The students frequently consult on classroom tasks with the teacher	35	75
		c. The students enjoy completing writing tasks more frequently	55	90
		d. The students complete the writing tasks on time	45	85

The total number of students involved in this research was 24, which was the actual number of class participants. They were identified individually in terms of their classroom motivation and participation through four criteria. As shown in Table 4, the results of the observations show a significant increase in participation and motivation indicators from cycle 1 to cycle 2.

3.1.3.3 EVALUATION RESULTS

Evaluation was conducted at the last (fourth) meeting of each cycle, which means that there were two evaluation sessions during the research, which were used to establish the real improvement in students' English writing skills after they had been treated. The results are shown in Tables 5 and 6.

3.1.3.4 REFLECTION

The last step in classroom action research is reflection, which is the stage where the researcher and the collaborating teacher review previous activities and the result of the tests for evaluation purposes. Reflection is directed to review all teaching activities the teacher did in the classroom.

Table 5. Results of evaluation of cycle 1.

No. of students	Evaluated writing components					Average of gained score
	Mechanics	Language use	Vocabulary	Organization	Content	
24	50.0%	54.16%	50.0%	37.5%	45.8%	63.75 45.8%

Table 6. Results of evaluation of cycle 2.

No. of students	Evaluated writing components					Average of gained score
	Mechanics	Language use	Vocabulary	Organization	Content	
24	95.83%	91.66%	95.83%	87.5%	87.5%	72.91 91.66%

Based on the results of reflection on cycle 1, it was found that: 1) the teacher's teaching performance was not well conducted, such that it was possible that it contributed to students' low classroom participation and motivation. As a result, the students' achievements in writing skills did not improve significantly, showing: 2) the students' classroom participation and learning motivation was categorized as low; 3) the students' achievement in writing skills was not improved significantly. This means that the minimum criterion for success wasn't achieved and, inconsequence, the researcher and the collaborating teacher decided to continue the research action through a second cycle.

3.2 *Discussion*

Our discussion covers three main areas, detailed in the following subsections.

3.2.1 *Teacher's Teaching Preparation and Performance*

Based on the results of observation, it was found that there was not any particular changes the collaborating teacher and the researcher did in the planning phases. Meanwhile, in terms of the teacher's teaching performance, the researcher and the collaborating teacher did make some changes in the teaching step. In the first cycle, the researcher missed out some steps in pre-activity, only implementing two of the four planned teaching steps, or 50%. In the while-activity, of 13 planned teaching steps, the researcher implemented six steps or 53.3%, and of six planned teaching steps in post-activity, the researcher implemented two of them, or 33.3%. However, in the second cycle, the researcher implemented all of the planned teaching steps in each of the pre-, while-, and post-activities.

The core of the teaching is in the while-activities, so the researcher is required to be more detailed in undertaking teaching steps here. While-activity is the place in which the researcher describes the steps involved in writing, how to write, and how to construct a recount text. In addition, during the while-activities the researcher provided more chances to students to ask clarification questions. What the researcher did in their teaching was in accordance with Fauziati (2010, p. 53), who stated that to teach genre is not merely to emphasize text itself but should be integrated with teaching writing to students.

3.2.2 *Students' Learning Motivation and Classroom Participation*

The observations the collaborating teacher and researcher made during lessons found changes in each of the four components of students' classroom participation and learning motivation that describe what the students should fulfill in order to indicate significant progress in these behaviors. The findings are explained as follows: a) the improvement in students' attention to teachers' instruction/explanation showed improvement of around 45%, from 45% in cycle 1 to 90% in cycle 2; b) the frequency in asking clarification questions also showed improvement of around 45%, from 25% in cycle 1 to 70% in cycle 2; c) the students' classroom collaboration also showed improvement of around 45%, from 40% in cycle 1 to 85% in cycle 2; d) the students' readiness to provide help to classmates demonstrated improvement of around 60%, from 15% in cycle 1 to 75% in cycle 2.

Similar improvements were seen in students' learning motivation: a) the students' punctuality presence showed an improvement of around 25%, from 65% in cycle 1 to 90% in cycle 2; b) the frequency of students' consultation in the classroom improved by around 40%, from 35% in cycle to 75 in cycle 2; c) the students' pleasure in completing classroom tasks showed improvement of around 35%, from 55% in cycle 1 to 90% in cycle 2; d) students' punctuality in completing classroom tasks showed an improvement of around 40%, from 45% in cycle 1 to 85% in cycle 2.

Based on the findings above, it is very clear that students' classroom participation and learning motivation improved significantly from cycle 1 to cycle 2, because each observed component showed an improvement beyond the criteria for success: 65. These findings support previous research conducted by Higgins, Dewhurst and Watkins (2012), who found that field trips engage "and even entertain students, helping to make the educational experience more enjoyable and... more memorable and more sociologically meaningful", as well as increasing the motivation of students in relation to their subject matter. It is clear that a field trip or field learning experience contributes to an increase in students' motivation to participate in the classroom activities.

3.2.3 *The Result of Evaluation of Students' Writing Skills*

There are two forms of evaluation employed to evaluate the students' writing skills in this research: the achievement score divided by the maximum score and multiplied by one hundred, and the mean score derived by calculating the aggregate scores gained by the students divided by the number of students and adapted to writing skills evaluation components, as suggested by Jacobs (1981).

The results of evaluation in cycle 1 are as follows: the average score gained by the students from the five writing components evaluated was 63.75. This is obtained by accumulating the total scores gained by the 24 students and dividing by the number of students. Meanwhile, the class achievement was only 45.8%. This percentage is derived by counting the number of successful students (the students who met or exceeded the critieria for success score) divided by the actual number of students, multiplied by one hundred.

By contrast, the results of the evaluation in cycle 2 show a significant improvement in student's English writing skills. The average score gained in cycle 2 was 72.91, and the classroom achievement was 91.66%. The way the researcher and the collaborating teacher determined this average score and percentage is similar to the approach of cycle 1. It can be said that students' writing skills improved well from cycle 1 to cycle 2. The difference in average score is around 9 points, and the scale of percentage change in class achievement is about 46%.

4 CONCLUSION

The outcome of the research can be summarized as follows. This research focused on investigating the strategy for teaching English writing skills by writing recount text, and students' achievements in English writing skills as a result. Observations were directed to the three main areas of this that relate to one another: the teacher's teaching preparation and performance, the students' classroom participation and learning motivation, and the improvement in students' writing skills.

We found that there was a significant improvement in the teacher's teaching performance from cycle 1 to cycle 2. The difference lies in the scale of percentage of teaching steps completion. The findings for students' classroom participation and learning motivation also showed significant improvement from cycle 1 to cycle 2. Again, the difference lies in the scale of the percentage for each observed component. The improvement in students' achievement from cycle 1 to cycle 2 is as follows: the average scale of students' score gains between cycle 1 and cycle 2 is about 9 points, from 63.75 in cycle 1 to 72.91 in cycle 2. Meanwhile, the difference in the percentage score for classroom achievement is almost 46%, going from 45.8% in cycle 1 to 91.66% in cycle 2.

Based on the results of the observation and evaluation done in this research, it can be said that the field learning experience strategy is effective in improving students' writing skills through the writing of recount text.

REFERENCES

Anderson, M. & Anderson, K. (1997). *Text types in English.* South Melbourne, Australia: Macmillan Education.

Brown, H.D. (2001). *Teaching by principles: An interactive approach to language pedagogy* (2nd ed.). White Plains, NY: Longman.

Bukian, P.A. (2004) Speaking Skill Teaching Method at Grade VI of Bukungkulan Elementary School Sawan District of Buleleng Regency (Unpublished Under Graduate Thesis. Singaraja State University

Celce-Murcia, M. (Ed.). (2001). *Teaching English as a second or foreign language* (3rd ed.). Boston, MA: Heinle & Heinle.

Clark, I.L. (2003). *Concept in composition: Theory and practice in the teaching of writing.* Mahwah, NJ: Lawrence Erlbaum Associates.

Fauziati, E. (2010). *Teaching English as a Foreign Language (TEFL).* Surakarta, Indonesia: PT Era Pustaka Utama.

Finchum, W.M. (2013). *How can teachers and students prepare for effective field trips to historic sites and museums?* (Doctoral dissertation, University of Tennessee, Knoxville, TN). Retrieved from https://trace.tennessee.edu/utk_graddiss/2569/

Flickr. (2013). *Recount Text: Contoh Pendek dan Generic Structure.*

Higgins, N., Dewhurst, E. & Watkins, L. (2012). Field trips as short-term experiential learning activities in legal education. *The Law Teacher, 46*(2), 165–178. doi:10.1080/03069400.2012.681231

Jacobs, H.L. (1981). *Testing ESL composition: A practical approach.* Rowley, MA: Newbury House.

Jacobsen, A.D, Eggen, P. & Kauchak, D. (2009). *Methods for teaching.* Yogyakarta, Indonesia: Pustaka Pelajar.

Kemmis, S. & McTaggart, R. (1988). *The action research planner.* Waurn Ponds, Australia: Deakin University Press.

McNiff, J. (1992). *Action research: Principle and practice.* New York, NY: Chapman & Hall.

Meiranti, R. (2012). Improving students' writing skills through field trip method. *English Review: Journal of English Education, 1*(1), 89–96.

Nunan, D. (1989). *Classroom action research.* New York, NY: Prentice Hall.

Orion, N.& Hofstein, A. (1994). *Factors that influence learning during a scientific field trip in a natural environment.* Journal of research in science teaching 31 (10) 1097-1119

Raimes, A. (1983). *Techniques in teaching writing.* Oxford, UK: Oxford University Press.

Syahmadi, H. (2013). The 2003 Curriculum Exploration for English Teachers. Bandung Indonesia: CV Adoya Mitra Sejahtera.

TVET Towards Industrial Revolution 4.0– Hazirah Noh@Seth et al. (eds)
© 2020 Taylor & Francis Group, London, ISBN 978-0-367-24273-2

Learning prospects for bioentrepreneurship in Indonesia: A study in junior and senior high schools

Tumisem & Epriliana Dewi
Department of Teacher Training and Education, Muhammadiyah University of Purwokerto, Purwokerto, Indonesia

ABSTRACT: Many students graduating from junior high school and senior high school in Indonesia do not continue their education to high school (university). This condition causes unemployment in Indonesia is very high This unemployment is dominated by junior and senior high school graduates.. Thus, the Indonesian government must find a solution. At present, the development of technology in education and training in all countries is one of the focuses that can generate employment. This activity can be started early, from elementary schools, junior high schools, and senior high schools, by introducing entrepreneurship-based learning opportunities and entrepreneurial prospects. In order for learning to be more meaningful, entrepreneurship learning is provided through a system of practice in all fields of science that enable entrepreneurship. This learning system can be implemented in an integrated manner, both interdisciplinary or multidisciplinary, namely interdisciplinary: learning is done integrated with allied fields of science namely chemistry with biological sciences, while multidisciplinary learning can be integrated between economics, chemistry and biological sciences. Interdisciplinary integration of science in Indonesia can be carried out by combining biotechnology material with entrepreneurial material, known as bioentrepreneurship. Biotechnology learning material includes biology and its applications, and entrepreneurial material includes the production process from the application of biological theory and marketing systems. This study analyzes aspects of production creativity assessed from the ability of junior and senior high school students in formulating, planning and producing (product results).. The implementation of the study used an experimental method (control group design) with a project system. Experimental material in this study is integrated with biological material, namely: the use of microorganisms, plant growth and development and biological use of waste.. The research data were collected through observation, tests and questionnaires. The results of the study illustrate the increasing understanding of junior and senior high school students of biological material, biology process applications and biotechnology-based entrepreneurship processes, as well as the entrepreneurial system, which includes production management, packaging and marketing. Increased understanding and ability of students with a score of three is very good. Thus, it can be concluded that bioentrepreneurship learning has very good prospects for fostering entrepreneurial spirit and has good prospects for developing business units in schools that can make income for schools in Indonesia, thereby reducing unemployment.

1 INTRODUCTION

The unemployment rate in Indonesia is very high. The unemployment rate from junior high schools is 5.18% and senior high schools is 7.19%, universities is 6.31% and elementary school graduates is 2.67% (Andri, 2018). This occurs because the amount of job opportunities in Indonesia is less than the amount of labor generated each year. A high unemployment rate in a country will lead to an increase in poverty levels. The solution is that the government should create new job opportunities, improve entrepreneurship skills, open up entrepreneur programs and provide facilities to such programs from junior high schools to senior high schools. The learning of entrepreneurship in Indonesia can be integrated with every subject in school. This learning demands the competence of a teacher in every subject area. Therefore, so that the learning of entrepreneurship should be integrated in each subject, a learning curriculum should be established, based on the analysis of the possibility of doing entrepreneurship in each subject area. This enables the learning of integrated entrepreneurship to be implemented in a sustainable manner, so that it will instill a spirit of entrepreneurship from an early age. This condition is expected to minimize unemployment and create new motivation to open up new jobs.

One type of entrepreneurship that can be applied in junior high schools and senior high schools is bioentrepreneurship. Bioentrepreneurship can include a variety of microorganism and biotechnology applications in accordance with everyday people's needs.. The application of technology in biology is often known as biotechnology. Communities in Indonesia are more familiar with conventional biotechnology than modern biotechnology. This is because conventional biotechnology only requires simple equipment that is easily provided and made by the community itself.. However, they are using conventional biotechnology to produce foods such as soy sauce, bread, and so on. The study of biotechnology is an excellent opportunity to develop the basics of biological entrepreneurship (bioentrepreneur or bioentrepreneurship). This learning can be done with a scientific approach that involves project-based learning, discovery or inquiry, and the laboratory. Therefore, the learning of bioentrepreneurship in both junior and senior high schools can also be used to develop the skills of the science process (Tumisem et al., 2017).

The basic idea of bioentrepreneurship is the discovery of several benefits from biology that can generate entrepreneurial opportunities and can increase the economic value of the community, so that the application of bioentrepeneurship enables it to provide employment opportunities. Aspects in bioentrepreneurship learning include: creativity, marketing and production

The marketing system in bioentrepreneurship learning is done through cooperative schools, web schools and collaboration between schools. Thus, bioentrepreneurship can be used to exploit and develop life science innovation in schools and improve science process skills, improve entrepreneurial spirit, and also open up entrepreneurial opportunities (Tumisem et al., 2018). The learning of bioentrepreneurship is learning that relates directly to real objects or phenomena around the student's life. In this study, we produce a product as a result of the process of science, so that students can learn the process of processing a material into a useful product that has economic value and market opportunities. Learning is designed and executed from objects or phenomena that can produce products of high economic value and market opportunities, which can lead to entrepreneurial processes (Epriliana, 2017).

2 PROBLEM BACKGROUND

The development of science and technology, which has very quickly impacted various aspects of life, include education from the basic level to higher education. Along with these developments, the school burden is increasingly heavy and complex. One aspect is that every school is required to prepare graduates who are able to compete in the world of work and produce graduates who have soft and hard skills. Learning activities, at the secondary education level, are much oriented toward learning that develops independence, creativity and innovation. Thus, graduates from secondary education are able to find work and can be entrepreneurial, oriented to individual needs and community needs. Therefore, learning at the secondary education level can be done with an entrepreneurial orientation. Entrepreneurial-oriented learning provides opportunities for graduates to develop critical thinking, creativity and innovation (Fadil, 2017).

3 RESEARCH METHODS

The study was conducted in ten junior and ten senior high schools for one semester with project-based learning. Each class in each school was grouped into ten groups. Each group consisted of five students. In each class there were four or five groups. Each group was assigned a different task. Group one served as a producer, group two served as production manager, group three served as packaging and product design, group four served as a seller and product promotion, and group five served as a marketing manager.

Performance evaluation of each group was done through observation and a questionnaire. Observation and questionnaire indicators included production capability, cooperation, accuracy, work discipline, ability to design and promote, marketing and managerial skills. Observation sheets and questionnaires have three score categories: score 3 (very good), 2 (good), and 1 (poor). The indicators in the questionnaire also include: the level of difficulty in production, packaging, marketing, entrepreneurial motivation and production creativity. Each indicator has three score categories: score 3 (very difficult), 2 (moderate), and 1 (easy). The test was conducted to obtain data on students' understanding of biological material and entrepreneurial material in the field of biology.

4 RESULTS AND DISCUSSION

In the first project, the production capacity of students from both junior and senior high schools had a score of 2 (good). Production capability was very good (score of 3). This condition occurs after the experiment runs three times. This occurred because students already know and understand the series of production processes based on experience, opinions of other groups, and references via the Internet. Good production results during bioentrepreneurship learning can illustrate that students have been able to work together well and show good work discipline, and enable them to be able to carry out entrepreneurial activities. Bioentrepreneurship learning and product learning outcomes are illustrated in Figure 1.

All groups had good packing ability (score 3). This condition occurs after the experiment runs four times. This was because students learn through repetition, which is followed by the ability to reflect on work. This condition shows that the motivation of students in entrepreneurship activities is very high and has great prospects for the development of entrepreneurship in Indonesia, starting from junior high and senior high schools.

The ability of students to market products and managerial abilities during bioentrepreneurship learning still has a low score (score 1), although the experimental activities are repeated up to six times. This shows that students experience difficulties in marketing and managerial

Figure 1. Production process.

aspects. Therefore teachers and school staff help to market student production result through the school web. This difficulty can be overcome if all school elements play an active role in assisting and facilitating marketing and managerial activities such as product promotion, and collaboration between school cafeterias and shops or supermarkets around the school.

Student production creativity from each cycle increased with increasing entrepreneurial motivation. Production creativity was evaluated from elements such as the ability to formulate a product, plan production and produce (Figure 2). Student creativity in production increased when teachers provide examples of similar products. This shows that students have understood the entrepreneur process based of content biology and biotechnology, so as to improve the creative attitude of junior and senior high school students (Figure 3). Student talents will influence production creativity. Students who have creative production can create something new, rather than students who do not have creative productive talent. Production creativity can be developed through a learning base in bioentrepreneurship that directs students to experiment directly. This experiment enables students to be selective and productive in producing a product.

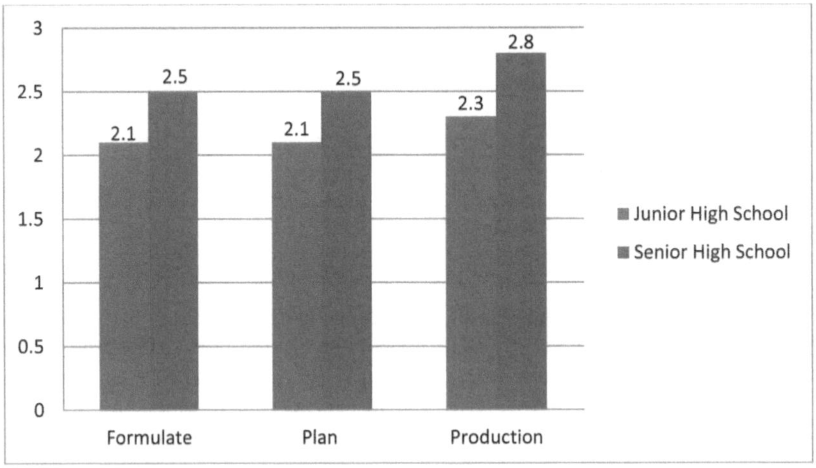

Figure 2. Student's creativity in six production cycles.

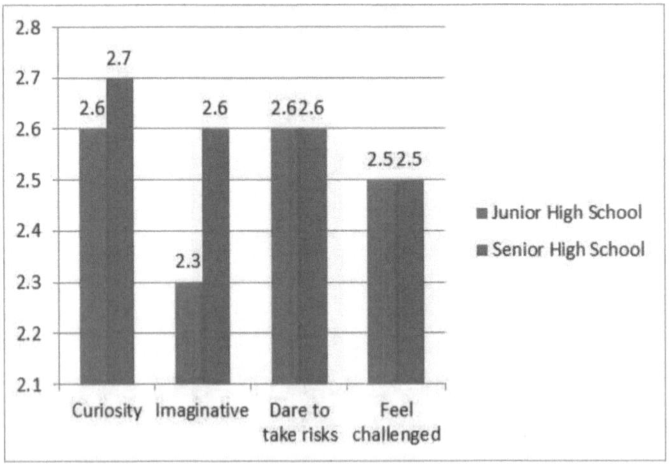

Figure 3. Student's creativity attitude in six production cycles.

5 CONCLUSIONS

Bioentrepreneurship learning, which is integrated in the learning of biology in junior and senior high schools, can improve entrepreneurial skills in a sustainable manner. This condition has very good prospects to reduce unemployment in Indonesia. This can happen because the alumni of junior high school and senior high school students who do not get a job in an agency, either government agencies or non-governmental institutions, can become an entrepreneur. This will have an impact on the creation of new job opportunities.

ACKNOWLEDGMENTS

Our thanks go to the Ministry of Research, Technology and Higher Education who have provided research funding from 2015 to 2017. We would also like to thank all schools that have coordinated the implementation of research for three years continuously, Muhammadiyah University of Purwokerto, which has provided funds to participate in conferences, and the third TVETIC conference that has collaborated to publish our article in the Journal of Technical Education and Training.

REFERENCES

Andri, D.P. (2018). *Unemployment from most primary school graduates, most graduate high. school vocational.* Retrieved from https://ekonomi.kompas.com/read/2018/05/07/155718426/pengangguran-dari-lulusan-sd-paling-sedikit-terbanyak-lulusan-smk

Epriliana, D. (2017). *Learning based on bioentrepreneurship. A scientific approach to the ability of creative thinking and attitude creativity students* (Thesis, Faculty of Biology, Muhammadiyah University of Purwokerto, Banyumas, Indonesia).

Fadil, A. (2017). *Critical thinking ability of students in the Xi class of Muhammadiyah Senior High School in Purwokerto between bio-entrepreneurship based learning with practicum learning* (Thesis, Faculty of Biology, Muhammadiyah University of Purwokerto, Banyumas, Indonesia).

Tumisem, Mufida.N., & Arief, H. (2017). Learning implementation strategy that based on bio-entrepreneurship in senior high school. In *Proceedings of The 1st International Conference of Research and Community Service (IRECOMS), Muhammadiyah University of Purwokerto, 2016.*

Tumisem, Arief H., Epriliana, D & Rostaman. (2018). Introduction of entrepreneurship program through biology learning at senior high school. In *Proceedings of The 2nd International Conference of Research and Community Service (IRECOMS), Muhammadiyah University of Purwokerto, 2017.*

TVET Towards Industrial Revolution 4.0– Hazirah Noh@Seth et al. (eds)
© 2020 Taylor & Francis Group, London, ISBN 978-0-367-24273-2

Training package development in bottle glass forming using the blow and blow process for glass forming operators

S. Deewanichsakul & B. Sramoon
Rajamangala University of Technology Thanyaburi, Khlong Luang, Pathum Thani, Thailand

ABSTRACT: The objective of this research was to develop a training package in bottle glass forming using the blow and blow process for glass forming operators. The researcher used a constructed training package to test a purposive sampling group of 25 people who are the glass forming operators in the Ayutthaya Glass Limited Company. The efficiency of the training package was analyzed by E1/E and a *t*-test for comparison of the result of trainees' learning achievement. Before conducting the training, a pretest was provided to the testing group or trainees for testing pre-knowledge. During the training program, trainees did an exercise to evaluate the training program. After the training completion, a post-test was provided to the trainees to evaluate their achievements. The exercise score and post-test were calculated in order to measure the effectiveness of the training package. The result of the evaluation shows that the training package has an effectiveness of 87.73/84.23. This is higher than the expected standard of 80/80. The analysis of the result reveals that the difference in the test scores taken before and after using the training package brought higher capacity to the trainees, which was statistically significant at the 0.01 level.

1 INTRODUCTION

Currently, the industrial sector plays an important role in Thailand's economic development. The ratio of industrial product value to gross domestic product and the ratio of industrial product export value to gross exports have increased rapidly. However, the manufacturing sector remains highly reliant on import technologies, which means that the sector still needs to emphasize the advantage of low-cost unskilled labor and lower total cost. In addition, development in this sector is based on value added not value creation. Thailand's productivity, which can lead to value creation, remains inferior because the country has lacked knowledge accumulation to develop its potential in knowledge cumulation. Additionally, the country has lacked a synergy between innovation, knowledge and technology. Therefore, Thailand needs to adapt from emphasizing traditional advantages toward emphasizing value creation via innovation, knowledge and technology to correspond with current challenges, risks, opportunities and threats, which can elevate the country's potential for worldwide competition (Office of Industrial Economics, 2011).

The glass manufacturing industry is an industry that supports other domestic industries like food, cosmetics and the pharmaceutical industry with a Thai baht (THB) marketing value of approximately 16,000 million. Around 90% of glass containers in Thailand are bottles, with a manufacturing rate of around 2.4 million tons (10,000 million bottles) a year (Small and Medium Industrial Institute, 2011). However, glass manufacturing is a specialized industry which relies on imported advanced technology. Glass bottle production comprises seven steps, namely: 1) material preparation; 2) mixing; 3) melting; 4) forming; 5) annealing; 6) inspection; 7) packaging (Bangkok Glass Company Limited, n.d.). In the forming stage, there are three processes, namely: 1) blow and blow; 2) press and blow; 3) narrow-neck press and blow, with each process being suitable for different kinds of bottle.

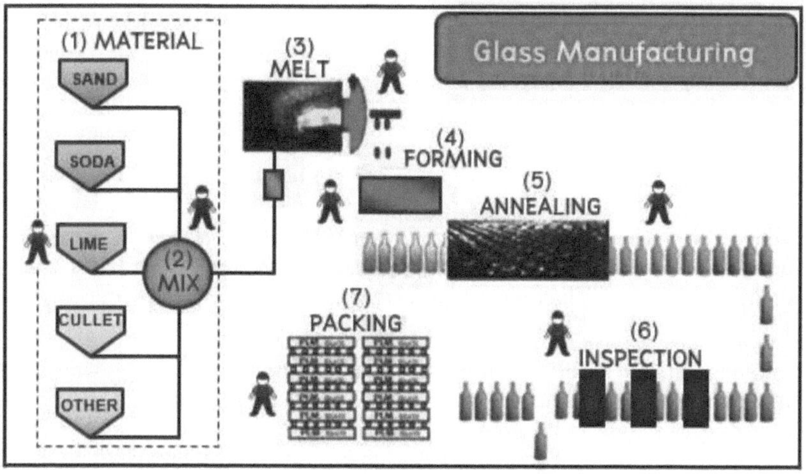

Figure 1. Glass manufacturing (Bangkok Glass Company Limited, n.d.).

Therefore, it is essential that glass forming operators understand device operation mechanisms and forming processes. A survey of glass forming operators' training needs found that training in the blow and blow process was the most needed, as the process comprises many steps and there are important details in each step that the operators need to know and understand as a basis to operate the forming and solve problems that occur. (Deewanichsakul & Wangworawong, 2014). A study of operator development found that most training was done via knowledge passing from senior operators, which resulted in each operator learning from different sources. The passed-on knowledge was not compiled and synthesized in a standard handbook that can be applied to operator training. Unfortunately, glass bottle manufacturing is a specialized industry that relies on imported automatic advanced technology; domestic education institutes and government agents cannot offer knowledge and advice to the section (Boontositrakool et al., 2017). Therefore, a standard training package in glass forming has been absent. Gathering knowledge and information from experts in this area and applying the data to a training package which can supply to the operators would be a good way to train them. The operators can apply the knowledge to the operation as a guideline for basic problem analysis and solving. The training package can be a systematic training tool for operators, which corresponds to a study that shows that a training package is a systematic tool and guideline for training both trainer and trainee. The package must be comprised of procedures, content, media, activities, experience and evaluation.

If the training issue is not solved, it will affect the operators' capacity. Consequently, researchers are interested in training package development for the blow and blow forming process and applying the package to train operators who, in turn, can use the knowledge to operate the process correctly and safely, which can increase manufacturing capacity.

2 STATEMENT OF THE PROBLEM

Glass container manufacturing is a specialized industry where the tacit knowledge about forming has been passed on individually, which causes difficulty in operator training and development. The research team conducted an opinion survey in an organization and found that they were interested in staff development and realized the importance of glass forming process development. As a result, the team synthesized the knowledge to develop the glass container manufacturing capacity of the industry.

3 OBJECTIVE

To develop a training package regarding bottle glass forming using the blow and blow process for glass forming operators.

4 METHODOLOGY

The design involved an experimental research approach, which aimed to develop a training package regarding the blow and blow forming process for the operators and to determine the training package's efficiency. The research procedures were:

(a) *Information gathering*
  (i) Studying detailed information about the blow and blow process, which is based on the expected capabilities that the trainees need to demonstrate after the training, according to the objectives.
  (ii) Deciding the training content via document studying, manual research and expert interviews. The information was analyzed according to the framework in order to decide the content.
  (iii) The content was evaluated by seven experts in glass bottle manufacturing. The experts' comments were applied to improve the content.

(b) *Training package development*
  (i) The instrument in the study was the training package on the blow and blow forming process. The instrument development process was as follows:
    (a) Setting behavioral objectives which emphasize trainees' behavior change. The areas of target behaviors were recalling knowledge, applying knowledge and transferring knowledge.
    (b) Training package development which comprised a trainer handbook, content, presentation, magnetic pictures, exercises and examination.
    (c) The training package was evaluated by the seven experts based on the following details:
      (i) Training package suitability was evaluated by the experts. Most of them agreed that the trainer handbook, the content, the media, the exercises and the examination were suitable for the training.
      (ii) Consistency of the examination was evaluated by the experts via *Index of Consistency*: IOC, which indicated that the consistency value was between 0.71–1.00, which was higher than standard (0.5).
    (d) A pilot test was launched for the package. The test was applied to a sample group of five glass forming operators. The test aimed to identify any flaws in training, content, media and language for correction before the package was used in real training.
    (e) After correction, the complete package was acquired and was ready to use in training.

(c) *Sample group*
  (i) The sample group was 25 glass forming operators who worked in the Ayutthaya Glass Limited Company. They were selected by purposive sampling.

(d) *Experiment and data collection*
  (i) The data collection process to determine the efficiency of the package was as follows:
    (a) Informed the trainees about the objectives of the training.
    (b) Launched a pretest to determine trainees' base knowledge.
    (c) Launched the training with the package, which was conducted by a foreman who had been trained in the teaching techniques. After each session finished, the trainee worked on an exercise to measure progression.
    (d) After completing the training, the trainees were tested using a post-test.
    (e) The scores from the exercise and the post-test were analyzed to determine the efficiency of the training package.

5 METHOD OF DATA ANALYSIS

The data were analyzed by means and standard deviations, which were compared with Best's standard mean (Best, 1977, p. 296), to determine the efficiency of the package. A *t*-test was also calculated.

6 RESULTS

The data were analyzed and yielded the results described in the following subsections.

6.1 *Training package efficiency*

Table 1 shows that the mean score of 25 trainees was 26.32 or 87.73%, and their post-test mean score was 29.48 or 84.23%.

Table 1. The training package efficiency analysis result.

	N	ΣX	$\overline{X}$	%
Exercise score (total 30)	25	658	26.32	87.73
Post-test score (total 35)	25	737	29.48	84.23

6.2 *Progression analysis*

Table 2 shows that the post-test score was significantly higher than in the pretest ($p < 0.01$).

Table 2. Difference analysis between before and after the training.

	N	ΣX	ΣD	ΣD^2	t
Before training	25	528	209	1839	21.38**
After training	25	737			

6.3 *Satisfaction analysis result*

Table 3 shows that the participants were highly satisfied with the training (= 4.00). When considering the various aspects, the most satisfying aspect was the media (= 4.41); second was the content (= 4.01), third was the trainer (= 3.85) and fourth was the administration (= 3.74). Overall, the results indicated that the participants were highly satisfied with the training.

Table 3. Trainees' satisfaction score.

No.	Aspect	$\overline{X}$	S.D.	Satisfaction level
1	Content	4.01	0.58	High
2	Media	4.41	0.49	High
3	Trainer	3.85	0.69	High
4	Administration	3.74	0.53	High
Total		4.00	0.57	High

7 FINDINGS OF THE STUDY

The training package efficiency analysis yielded the following results:

(a) The mean exercise score of the participants (25 operators) was 26.32 or 87.73%, while their mean examination score was 29.48 or 84.23%, which meant that the training package efficiency score was 87.73/84.23.
(b) For their progression, the post-test score was significantly higher than the pretest score (at.01 level of significance), which indicated that the trainees progressed.
(c) Regarding satisfaction, the mean score was 4.00, which meant that the trainees were highly satisfied with the training.

8 DISCUSSION

This study was conducted to develop a training package and determine its efficiency. The package was developed and applied to 25 glass forming operators who worked in the Ayutthaya Glass Limited Company. The data collection began with giving information about the objectives and then launching the pretest. The operators participated in the training session with the foreman who had been trained in the teaching technique before. After the training, the participants were tested again. The scores were analyzed to determine the training package efficiency. The results indicated that the efficiency score was 87.73/84.23, which shows that the package can be used to train efficiently. Furthermore, the mean post-test score was significantly higher (at the .01 level) than the mean pretest score, which means that the participants progressed. There was a variety of media in the package, such as documents, animation media and, especially, magnetic pictures. The magnetic pictures were very suitable for training in the blow and blow forming procedure, because there were a total of 37 working steps of mechanics and molds in the process and the magnetic pictures allowed the trainees to put each step in order, which made them truly understand the overall process. The result corresponded with the process of active learning as described by Srichancheun (2011), who described the meaning of active learning as 'the learning that besides reading and listening, students should have more opportunity to practice, which leads to thinking, analyzing, synthesizing and evaluating; the procedure would transform the learner to be the knowledge creator'. The research result shows that the developed training package can be used to train the glass forming operators to have knowledge of the process of bottle glass forming in using the blow and blow process effectively. In addition, they can apply knowledge, which gained from the training as a guideline to solve the problem of defects in the production process correctly and the same standard which effected to the establishment has higher productivity.

Regarding satisfaction, the participants were highly satisfied with the training but this may be because they needed the content in their work. The result of this study corresponds with that of Deewanichsakul and Wangworawong (2014), who studied their training package in glass bottle forming, which gained a 83.33/80.21 efficiency score. Additionally, the result corresponds with that of Boontositrakool et al. (2017) regarding training package development to solve a problem in the blow and blow forming process and which achieved a 82.20/81.06 efficiency score. The score of all studies was above the standard score (80/80), which meant they can be applied to training efficiently.

9 CONCLUSION

The results indicated that the training package was efficient and could be used in real training. The glass forming operators can apply the knowledge which gained from the training to use as a guideline in problem solving. Consequently, they will be able to analyze the cause and solve problems from the production process more accurately and quickly, leading to higher production efficiency.

10 RECOMMENDATIONS

In the future, another training package, in an active learning framework, should be developed for training in glass container manufacturing. Furthermore, entrepreneurs should encourage their employees to train in active learning programs, which can assist them to develop systematically.

REFERENCES

Bangkok Glass Company Limited. (n.d.). *Training manual for glass forming operators.* Pathum Thani, Thailand: Technical Training Department, Bangkok Glass Company Limited.

Best, J.W. (1977). *Research in education* (4th ed.). Englewood Cliffs, NJ: Prentice Hall.

Boontositrakool, K., Deewanichsakul, S. & Wannaharnon, A. (2017). The training package development on cause and problem resolving of thin body bottle for blow and blow process. *Journal of Technical Education Development, 29*(100), 80–85.

Deewanichsakul, S. & Wangworawong, W. (2014). The training package development to enhance the glass production competency for glass forming operators. *Princess of Naradhiwas University Journal, 6*(2), 83–91.

Office of Industrial Economics. (2011). *National Industrial Development Master Plan.* Bangkok: Ministry of Industry.

Small and Medium Industrial Institute. (2011). *Reporting of analysis for the capabilities of the ASEAN economic community.* Bangkok, Thailand: Glass Industrial.

Srichancheun, M. (2011). *The teaching a large group in Gsoc 2101 the development community: Approach to learning by active learning and using e-learning to develop a model of teaching in higher education, learning and academic achievement of students.* Chiang Mai, Thailand: Faculty of Humanities and Social Sciences, Chiang Mai Rajabhat University.

TVET Towards Industrial Revolution 4.0– Hazirah Noh@Seth et al. (eds)
© 2020 Taylor & Francis Group, London, ISBN 978-0-367-24273-2

Enhancement of the Malaysian Qualification Framework for equivalence-checking via APEL

N.F.M.Mohd Amin & N. Kaprawi
Faculty of Technical and Vocational Education, Universiti Tun Hussein Onn Malaysia, Batu Pahat, Johor, Malaysia

ABSTRACT: The Malaysian Qualification Framework (MQF) is a measurement tool developed based on accepted and valid criteria at the national level. The results of this new MQF can be used to check for an equivalence through Accreditation of Prior Experiential Learning (APEL). APEL is a system that recognizes learning based on past experience, which is helpful for individuals with much experience but no opportunity to continue their studies. APEL will take on lifelong learning and apply it in everyday life. This study aimed to explore the constructs within the ideal framework for the improvement of its use in the prior knowledge-based recognition rubric (APEL). Some qualification frameworks are benchmarked: the European Qualification Framework (EQF), Australian Qualification Framework (AQF), South African Qualification Framework (SAQF), German Qualification Framework (DQF), New Zealand Qualification Framework (NZQF) and Scottish Qualification Framework (SCQF). The approach used in this study is qualitative. Data obtained in this study is through document analysis. This study can help in developing a more comprehensive and detailed eligibility framework for the intermediation between academic and vocational skills through APEL.

1 INTRODUCTION

The Malaysian Qualification Framework (MQF) is a measurement tool developed based on benchmarked international practice. It provides information on learning levels, learning outcomes and credit loads. In addition, the MQF also serves to provide educational facilities that can be used systematically and allow one to improve knowledge to higher education. Enhancement of the MQF will assist in the provision of a rubric that can be used to check for an equivalence through Accreditation of Prior Experiential Learning (APEL). APEL is an important element in informal learning recognition. Accreditation is a process of formal recognition, assessment and recognition of learning. Accreditation of Prior Learning (APL) is usually used to demonstrate various formal approaches as evidence that learning has already taken place. According to Mohd Tahir and Mustafa (2009), technical and vocational education can be used as one of the solutions in recognizing talented and skilled students. These are the ones that will help the country in generating the economy, fulfilling industry expectations and upholding a nation in the eyes of the world.

Hence, producing skilled graduates must be realized through the development of a system that is not merely academic-focused but gives more attention to Technical and Vocational Education and Training (TVET) (Malaysian Ministry of Education, 2015). In addition, Malaysia's qualification framework also plays a core role in improving education in the country. The formation of a national eligibility framework of a country should focus on learning outcomes. The learning outcomes are instrumental in providing a detailed description and are able to identify the difficulty of a feasibility level within the feasibility framework. A comprehensive qualifying framework is a result of concepts, values and views from training and education covering the entire framework of qualification itself. According to the European Union (2011), the description level should include the following:

i) complex diversification in the national qualification system;
ii) relevant diversity in workmanship;
iii) systematic differences between learning levels and illustrations of how knowledge, skills and competencies are acquired in the learning process;
iv) improvement of the description so that it becomes a benchmark for other countries.

Balasingam (2015) states that the MQF should be developed according to the same benchmark, despite different fields or disciplines. It should be made in detail and more clearly to illustrate the development of learning in terms of knowledge, specialized skills and generic skills, as well as the application of knowledge and skills at different levels. The MQF's description is still undetailed compared with the eligibility criteria of other countries such as Europe and Australia. However, the level of education in Malaysia has eight levels, similar to that in Europe. Furthermore, a complete description of each level is still missing. In the MQF there is a description of levels one to three but the learning outcomes for each of the levels in the MQF is not specifically mentioned. If this is refined, then the process of equivalence between the academic and vocational will be easier to implement.

APEL in Malaysia is increasingly being improved and recognized by the public. Currently, APEL has implemented accreditation for the experience of individuals. However, APEL still has no rubric to recognize the skills of students to extend to academia. This study is the starting point for assisting in the construction of a rubric of equivalence between skills and academia. With the ideal constructs, the descriptions of each stage will be formed more purely. This description is the second step before the implementation of equivalence between skills and academia.

2 OBJECTIVES

The objective of this study is to explore the constructs within the ideal framework for the strengthening of its use in the APEL rubric.

3 METHODOLOGY

This study uses a qualitative approach to explore constructs within an ideal eligibility framework for the strengthening of its use in the prior knowledge-based rubric recognition column (APEL). The data used in this study arise through document analysis. In general, document analysis involves of printed, published or depicted communications messages. The documents analyzed are the MQF, the European Qualification Framework (EQF), the Australian Qualification Framework (AQF), the South African Qualifications Framework (SAQF), the German Qualification Framework (DQF), the New Zealand Qualification Framework (NZQF) and the Scottish Qualification Framework (SCQF). The eligibility criteria for these countries were selected because of their stable framework and the implementation of a robust APEL system in their country. The process of analyzing the documents is as suggested by Braun and Clarke (2006). The five stages are carried out in analyzing these documents are:

i) Pre-analysis stage. All documents are read, reviewed, ranked and completed. All notes are labeled for type.
ii) Data packing stage. Data are transcribed in word processing or a textual analysis format using a computer.
iii) Data rating stage. Data are displayed in the form of tables, tally sheets and formulas.
iv) Data verification stage. All procedures that lead to conclusions are accurately stated.
v) Findings stage (including interpretation). Review and interpretation of the findings.

4 FINDINGS

Some qualification frameworks have fixed constructs, such as the EQF, AQF, DQF and the NZQF. Based on these frameworks, the constructs identified are knowledge, skills and competence. While the SAQF and the SCQF have several constructs, these constructs can be placed under three major constructs, as in other eligibility frameworks. For example, in the South African eligibility framework, the construct of knowledge may include the scope of knowledge and literacy knowledge.

In addition, methods and procedures, and problem solving, as well as context and systems can be incorporated into the skills construct. Some other constructs expressed in this qualification framework can also be classified in the competence constructs of: management of learning; ethics and professional practices; accessing, processing and managing information; producing and communicating of information; accountability. As a result, the benchmark for each qualification framework is collected, as shown in Table 1.

Based on the comparative document analysis of several implemented benchmarks from different countries, the following is a construct of the framework of eligibility for the enhancement of the MQF to facilitate the details of the description and develop of rubric equivalence-checking through APEL:

Knowledge
Skill
Competence
i. Professional skills and ethics
ii. Communication skills/ICT, calculation and entrepreneurship
iii. Autonomy, accountability and teamwork

5 DISCUSSION

According to Gudeva et al. (2012), a qualification framework is an instrument for the development, classification and recognition of skills, knowledge and competence based on certain stages. This affects whether or not you want to implement intermediation between the academic and the vocational. Therefore, this study can help in developing a more comprehensive and detailed eligibility framework.

The learning domain of the MQF (Malaysian Qualifications Agency, 2012) includes: knowledge; practical skills; social skills and responsibilities; values, attitudes and professionalism; communication skills, leadership and teamwork; problem-solving skills and scientific skills; information management skills and lifelong learning; and managing and entrepreneurial skills. All of these domains include the MQF but are not classified by domain. Further study should involve classifying the results of learning within the framework of feasibility based on the domains and constructs that have been studied.

According to the Malaysian Qualifications Agency (2012), APEL is a systematic process involving the identification, documentation and assessment of learning based on past experience, such as knowledge, skills and attitudes, to determine the level of an individual achieving the learning outcome desired to access a program of study. Therefore, the development of a main construct that needs to be applied within the MQF will help in this systematic process.

6 CONCLUSION

Overall, APEL is a new credential system in Malaysia and the main factor in producing this system is to encourage lifelong learning among our society today. To ensure a strong APEL implementation, the party responsible for implementing it should give priority to systematic and strategic work. Hence, in this era of globalization, educational organizations in Malaysia

Table 1. Qualification frameworks benchmarked.

	Africa (SAQF)	Australia (AQF)	Europe (EQF)	Germany (DQR)	New Zealand (NZQF)	Scotland (SCQF)
Knowledge	Scope of knowledge Knowledge literacy	Knowledge	Knowledge	Professional competence knowledge	Knowledge	Knowledge and understanding
Skills	Method and procedure Problem solving Context and system	Skills	Skills	Professional competence skills	Skills	Practice: applied knowledge, skills and understanding Communication, ICT and numeracy skills Autonomy
Competence	Ethics and professional practice Accessing, processing and managing information Producing and communicating of information Accountability Management of learning	Application of knowledge and skills	Competence	Personal competence (social competence and autonomy)	Application of knowledge & skills	Accountability and working with others Generic cognitive skills

play an important role in pushing for change to advance society and uphold Malaysia in the eyes of the world.

ACKNOWLEDGMENTS

This research was supported by a scholarship from the Ministry of Higher Education (MyBrain15) and Universiti Tun Hussein Onn Malaysia.

REFERENCES

Balasingam, U. (2015) *Malaysian qualification framework: A need to revisit*. University of Malaysia. Retrieved from http://eprints.um.edu.my/13545/1/MDP-99_.pdf

Braun, V. & Clarke, V. (2006). Using thematic analysis in psychology. *Qualitative Research in Psychology*, *3*(2), 77–101.

European Union. (2011). *Using learning outcomes: European Qualifications Framework Series: Note 4*. Luxembourg: Publications Office of the European Union. Retrieved from http://www.cedefop.europa.eu/files/Using_learning_outcomes.pdf

Gudeva, L.K., Dimova, V., Daskalovska, N. & Trajkova, F. (2012). Designing descriptors of learning outcomes for higher education qualification. *Procedia Social and Behavioral Sciences*, *46*, 1306–1311.

Jasmi, K. A. (2012). Data Collection Methodology in Qualitative Research. Qualitative Research Course Series 1 2012.

Malaysian Ministry of Education. (2015). *Malaysia education blueprint 2015–2025 (higher education)*. Putrajaya, Malaysia: Kementerian Pendidikan Malaysia.

Malaysian Qualifications Agency. (2012). *Guidelines to good practices: Accreditation of prior experiential learning*. Petaling Jaya, Malaysia: Agensi Kelayakan Malaysia.

Tahir, L. M., Mustafa, N. Q., & Yassin, M. H. M. (2009). TECHNICAL AND VOCATIONAL EDUCATION FOR SPECIAL NEEDS STUDENTS. *Journal of Educators & Education/Jurnal Pendidik dan Pendidikan*, *24*.

TVET Towards Industrial Revolution 4.0– Hazirah Noh@Seth et al. (eds)
© 2020 Taylor & Francis Group, London, ISBN 978-0-367-24273-2

Implementation of the 21st century learning approach among technical and vocational education trainee teachers

R.M. Zulkifli & M.A.Mohd Hussain
Universiti Pendidikan Sultan Idris, Tanjong Malim, Malaysia

N.H.Abd Wahid & N. Suhairom
Universiti Teknologi Malaysia, Skudai, Malaysia

R. Che Rus
Universiti Pendidikan Sultan Idris, Tanjong Malim, Malaysia

ABSTRACT: The purpose of this study is to identify the implementation of the 21st century learning approach among technical and vocational education trainee teachers as well as the challenge faced by trainee teachers in implementing the learning approach. A survey was conducted on 142 trainee teachers from Faculty of Technical and Vocational (FTV) of Universiti Pendidikan Sultan Idris (UPSI), Malaysia. A descriptive analysis involving percentage values, standard deviation, and mean were used to present the findings. The findings of the study indicate that 96% of the FTV students applied the 21st century learning approaches during their teaching practice period. The study also found that the developing of i-Think thinking map is the most popular 21st century learning activity implemented by respondents, while the exploration of creativity and innovation in students' group work was the most popular 21st century learning strategies being applied in respondents' classroom. However, the study indicated several barriers in the implementation of the 21st century teaching strategies and activities, particularly in regard of time and space constraints.

1 INTRODUCTION

The 21st century skills among the workforce are important in ensuring the fulfilment of the job market demand. Studies conducted in Malaysia have also proven the importance of the 21st century skills (Rohani, Hazri & Mohammad Zohir, 2017). As example in the manufacturing industry, employers in Malaysia have given priorities to hire workers with good thinking and interpersonal skills as well as those with good personal qualities (Rasul et al., 2009). In order to ensure that the Malaysia workforce are having sufficient 21st century skills, the integration of the 21st century skills elements in the field of technical and vocational education is seen as vital. One of the approach that have been implemented to integrate the 21st century skills in the education setting is through the 21st century learning approach. The 21st century learning approach refers to a list of strategies and activities that can be used in teaching and learning activities to provide students with opportunities to enhance their personal and interpersonal qualities (Malaysia Ministry of Education, 2016). The 21st century learning approach also refer to strategies to enhance the effectiveness of the learning sessions and at the same time, combine the 21st century skills with the content and teaching practices in schools (AACTA & the Partnership for 21st Century Skills, 2010). The purpose of implementing the 21st century learning approaches in classroom setting are very clear which is to provide students with essential knowledge, at the same time encourage the development of personal qualities, thinking and interpersonal skills among school students for their benefits in the future.

The 21st century learning approach has been well-received and implemented by the community of Universiti Pendidikan Sultan Idris (UPSI), which is the university that has produced the biggest number of education graduates in Malaysia. The implementation of the 21st century learning approach have been listed as one of the requirement in every UPSI's undergraduate and post-graduate courses. Hence, the UPSI's students have been exposed to the strategies and activities related to the 21st century learning approaches through their courses. In, addition, students receive a formal training to implement the 21st century learning approaches during their pedagogical classes as a requirement to go for practicum in the real school setting. As a pioneer education university in Malaysia, UPSI have high expectation that their students will implemented the 21st century learning approaches in their classroom during the teaching practice programme also after they become a permanent school teachers. Hence, an initiative has been taken to conduct this study in order to find out the implementation of the 21st century learning approaches among Faculty Technical and Vocational (FTV) students from UPSI during their teaching practice programme. The result of this study will be used to reflect the effectiveness of the training and encouragement given to FTV students regarding the implementation of 21st century learning approaches. Additionally, this study examines the obstacles and challenges faced by the trainee teachers in implementing the learning strategies of the 21st century so that the solutions can be plan and suggested for better practices.

2 21ST CENTURY LEARNING APPROACH

The 21st century learning approaches refers to a student-centered learning approach that encourage the development of students' communication, collaboration, critical and thinking skills, as well as creativity, values and ethical (Beagle, 2010; Malaysia Ministry of Education, 2014). The Ministry of Education Malaysia introduced the concept of The 21st century learning in 2014 in response to the need of developing 21st century skills among the citizens from the school age for the benefits of their future because those skills become important requirement to enter the job market (Malaysia Ministry of Education, 2016). The Malaysia Ministry of Education 21st century learning concept involve a list of active learning strategies designed to enhance the students' personal and skills development including: i) Higher order thinking skills, ii) communication, iii) self-reflection iv) collaborative, v) problem solving, vi) the use of technology, vii) creative thinking, and viii) critical thinking. The Malaysian 21st century learning concept also suggest seven students' activities that have been identified best to be implemented during the learning session which include: i) Development of I-think map, ii) gallery walk, iii) round table, iv) three stray one stay, v) problem solving, vi) think-pair-share, and vii) role play.

According to the Malaysia Ministry of education (2014), the introduction 21st century learning concept come with a goal to equip students with important life skills including communication, collaboration, and problem solving. Beside that, students that undergone the 21st century learning were expected be better prepared for assessment and examination since they frequently trained to deal with high order thinking situations, questions and issues. While, the ultimate goal is for students' early preparation to enter the job market since the current trend of employment prioritize candidates with good communication, critical thinking and problem solving skills.

Figure 1 shown below refers to a conceptual framework of the 21st century learning approaches modified from a list of 21st century skills learning strategies and activities suggested by the Malaysian Ministry of Education. By referring to the conceptual framework in Figure 1, this study was carried out to review the learning strategies and activities implemented by the FTV trainee teachers during their teaching practice in the real school setting. It also studied the constraints that arise in implementing the 21st century learning strategies and activities in order to improve the pedagogical training program at UPSI as well as teaching activities in school setting.

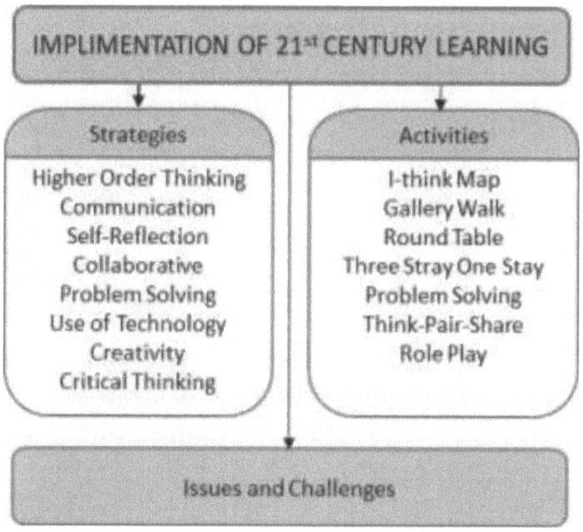

Figure 1. Conceptual framework of 21st century learning.

3 RESEARCH METHODOLOGY

This study uses a quantitative research method in the form of surveys. The researcher used a questionnaire developed with the aim to get responses from the trainee teachers regarding the implementation 21st century learning during their teaching practice programme. The five-level Likert Scale (Strongly Disagree – Strongly Agree) was used in the questionnaire to obtain feedbacks from the respondents. As the scope of the study involves only students from the FTV UPSI undergoing the teaching practice in 2017, all the students from the faculty were selected as respondents, with the total amount of 142 respondents. Those respondents come from three departments in FTV which are the Department of Agricultural Science, Department of Engineering Technology and Department of Family and Consumer Science. All these respondents have undergone teaching practice for six months at secondary schools and vocational Colleges and have been assigned to teach subjects related to technical and vocational education.

The researcher has obtained the validity of the research instrument before distributing it to the respondents. The instrument validity was performed by two experts with PhD qualifications in pedagogical areas and having more than 10 years of research experience. The validity of the instrument was intended to ensure accuracy of words, sentences, and language as well as content of the instrument (DeVellis, 2003). In order to ensure the reliability of the intrument, an initial pilot study was conducted by distributing 20 questionnaires to students of different faculties but underwent the teaching practice during the same session. The analysis of Alpha Cronbach coefficient value obtained for the sets of questions related to the implementation of 21st century learning was 0.719, whereas the set of questions pertaining to constraints faced by respondents was valued at 0.741. According to Mohd Majid Konting (2005), the research instrument has a high reliability value if the Alpha Cronbach coefficient value is 0.6 and is approaching the score of 1.0.

In this research, the reporting of the findings uses percentage scores, mean, mode and standard deviation. The mean range as suggested by Heir (2007) as in table 1 has been used to facilitate explanations and conclusions regarding the findings of the study.

Table 1. Interpretation on mean score value

Mean	Implementation level
1.00-1.99	Low
2.00-3.99	Moderate
4.00-5.00	High

4 RESEARCH FINDINGS AND DISCUSSIONS

4.1 *Implementation of 21st century learning activities by FTV trainee teachers*

The result of the study found that almost all of the trainee teachers (96.3% of 142 people), from the Faculty of Technical and Vocational, UPSI implemented the 21st century learning strategies and activities during their teaching practice at schools. Table 2 shows the activities performed according to popularity. The findings show that the development of i-Think thinking map is the most popular activity compared to other activities with 71.8% of the respondents carried out the activity. The gallery walk recorded as the second as second favorite activity with 68.3% of implementation, while the round table in the third position with 62.7% of implementation by respondents. The reason behind highly implementation of i-Think thinking map is because it is easy applied as it has eight different choices of interesting map (Nurhafizah, Roslinda, and Yusoff, 2015). This provides a wide choice to trainee teachers to apply them in their students' learning activities. The advantage is the i-Think map gives an interesting and creative learning experience to students at the same time enhance the cognitive stimulus in the forms of visual and graphic. (Nurhafizah, Roslinda, and Yusoff, 2015).

Table 2. Implementation of 21st century learning activities by FTV trainee teachers

21st century activities	Popularity	Yes(%)	No(%)
i-Think	1	71.8	28.2
Gallery Walk	2	68.3	31.7
Round Table	3	62.7	37.3
Three Stray One Stay	4	41.5	57.7
Problem Solving	5	41.5	57.7
Think-Pair-Share	6	35.2	64.8
Role Play	7	32.4	67.6
Others	10	12.0	88.0

4.2 *The 21st Century Learning Strategies Implemented by FTV Trainee Teachers*

Table 3 shows the analysis of data obtained from the questionnaire on the 21st century learning strategies that have been applied by trainee teachers during their teaching sessions. The results of the analysis show that most of the respondents highly implement all of the seven 21st century learning strategies listed in the questionnaire. The seventh strategy which is "The exploration of creativity and innovation in students' group work" had the highest mean score of 4.64. One factor that made this strategy popular among teachers is due to the interesting process of learning where students learn based on ideas and existing knowledge which are then structured by new ideas that go through the process of exploration and research (KSSR 2010).

Table 3. The mean value and standard deviation of 21st century learning strategies implemented by students

No	21st century learning strategies	Mean	Standard Deviation
1	Higher order thinking skill questions	4.18	0.67
2	Communication in groups	4.33	0.63
3	Self-reflection after teaching and learning sessions	4.21	0.74
4	Cross discussion between groups (collaborative)	4.48	0.40
5	Solving problems related to real life issues	4.30	0.70
6	Use of technology in completing tasks	4.08	0.70
7	Exploration of creativity and innovation through students' group work	4.64	0.31

4.3 *Challenges Faced by the FTV Trainee Teachers During the Implementation of the 21st Century Learning Approaches*

Table 4 shows the result of analysis regarding to the 10 challenges faced by the trainee teachers during the implementation of the 21st century learning approaches. Out of the 10 items, "time constraint" was identified as the main obstacle faced by the trainee teachers with the mean value of 4.04, followed by "lack of equipment and technology (3.89 mean value) and low confident level among students (3.86 mean value). Time constraints usually happen due to the extra time needed by teachers and students during the class transition. According to Ramlah (2018), during the class transition most of students and teachers have to move from one learning space to other, as example from classroom to laboratories or from workshop back to their classroom. Many school in Malaysia are big with various building and spaces which require extra time for students and teachers to move during the class transition. This movement during the class transition cause time to be wasted up to 15 minutes before the learning session takes place. The classroom transition process causes the learning session to start late and disrupts teachers' activity. Besides that, the respondents are unable to complete their teaching activities that have been planned in the daily lesson plan at the designated time.

Table 4. Mean value and standard deviation of the challenges faced by trainee teachers in implementing the 21st century learning approaches

No	Question	Mean	Standard Deviation	Rating
1	Time constraints	4.04	1.02	Strongly agree
2	Limitation of classroom space	3.68	1.18	Moderately agree
3	No cooperation from students	3.46	1.27	Moderately agree
4	Large number of students	3.54	1.27	Moderately agree
5	Obstructions in communication	3.68	1.11	Moderately agree
6	Language barrier	3.65	1.13	Moderately agree
7	Students unable to answer and think using the higher order thinking skills (KBAT)	3.77	1.02	Moderately agree
8	Students not interested in the learning activities	3.25	1.28	Moderately agree
9	Students with low confidence e.g. shy to respond	3.86	0.93	Moderately agree
10	Lack of equipment and technology	3.89	1.10	Moderately agree

5 CONCLUSION AND SUGGESTIONS

In conclusion, the 21st century learning approaches are diverse teaching and learning method aimed to make the teaching activity more interesting. This strategy also has the potential to develop the 21st century skills among students which have been seen as important in developing their personality and preparing them to enter the job market. The implementation of the 21st century learning approaches amongst trainee teachers at the Faculty of Technical and Vocational, UPSI is seen to be encouraging which indicate that the faculty and university manage to encourage their undergraduates FTV students to implement the 21st century learning strategies and activities in the real classroom setting. However, there are still some 21st century learning activities which are poorly executed by the trainee teachers, such as role-play and think-pair-share. Efforts should be made at the university level to enhance the confidence and skills of the future teachers to try and implement these activities. Other than that, the time constraints faced by the trainee teachers need to be given attention by schools' administrations so that the time of students and teachers is not wasted and the opportunity to implement 21st century learning approaches can be increased. At university level, initiatives through the pedagogical classes should be taken to train students to be able to implement the 21st century learning approaches within short and thigh period in order to assure it still can be done within the time constraint obstacle.

REFERENCES

AACTE & the Partnership for 21st Century Skills. (2010). *21st Century Knowledge and Skills in Educator Preparation*. Retrieved on February 2019 at http://www.p2.org/storage/documents/aacte_p21_whitepaper2010.pdf

DeVellis, R. . (2003). *Scale development theory and applications*. Thousand Oaks, California: Sage Publications.

Donald, B. (2010). The emergent information commons: Philosophy, models, and 21st century learning paradigms. *Journal of Library Administration*, 50(1): 7–26. DOI: 10.1080/01930820903422347

Kementerian Pendidikan Malaysia. (2014). *Kurikulum Abad Ke-21*. Bahagian Pembangunan Kurikulum, KPM

KSSR, Elemen Merentas Kurikulum. (2010). *Kreativiti dan inovasi*. Bahagian Pembangunan Kurikulum, Kementerian Pelajaran Malaysia.

Mohd Majid Konting. (2005). *Kaedah penyelidikan pendidikan*. Kuala Lumpur: Dewan Bahasa dan Pustaka.

Nurhafizah Zaidi, Roaslinda Rosli, dan Mohamed Yusoff Mohd Nor (2015). Aplikasi pemikiran i-Think dalam proses pengajaran dan pembelajaran matematik. Universiti Kebangsaan Malaysia.

Ramlah Muda. (2018). Pelaksanaan pengajaran dan pemudahcaraan abad ke-21 dalam kalang pelajar FTV semasa latihan mengajar (Unpublished undergraduate thesis). Universiti Pendidikan Sultan Idris, Malaysia.

Rasul, M.S., Ismail, M.Y., Ismail, n., Rajuddin, R. & Rauf, R.A. 2009. Aspek kemahiran 'employability' yang dikehendaki majikan industri pembuatan masa kini. *Jurnal Pendidikan Malaysia*, 34(2): 67–79.

Rohani Arba., Hazri Jamil., & Mohammad Zohir Ahmad (2017). Model Bersepadu Penerapan Kemahiran Abad ke-21 dalam Pengajaran dan Pembelajaran. *Malaysian Journal of Education (0126–6020)*, 42(1).

TVET Towards Industrial Revolution 4.0– Hazirah Noh@Seth et al. (eds)
© 2020 Taylor & Francis Group, London, ISBN 978-0-367-24273-2

Professional development needs of interim teachers in Malaysian vocational colleges

M.A.Mohd Hussain, N.S. Ibrahim, R.Mohd Zulkifli, A. Kamis, S. Mohamed & N.N. Mohd Imam Ma'arof
Universiti Pendidikan Sultan Idris, Tanjong Malim, Malaysia

ABSTRACT: This study was conducted to identify the challenges faced by interim teachers in Malaysian Vocational Colleges in their early teaching profession. A qualitative approach was used and the data were collected through interview sessions conducted with five interim teachers from two Malaysian Vocational Colleges. The framework of teaching by Charlotte Danielson was used as a guideline in studying the challenges faced by interim teachers in vocational colleges. The findings show that the biggest challenges faced by interim teachers were related to work stress and managing students' behavior, and discipline during teaching and learning activities. It is hoped that this study becomes a guideline for vocational college administrators, the Malaysian Institute of Teacher Education (IPGs), as well as the Technical and Vocational Education Division (BPTV) itself, in order to pave a better transition for the Malaysian Vocational Colleges interim teachers to become professional educators.

1 INTRODUCTION

The Technical and Vocational Education and Training (TVET) emphasizes developing knowledge and skills as a preparation for good employment opportunities for youngsters (Siti Maspah & Nor Azizah, 1995). TVET also plays an important role in preparing a capable workforce that is innovative, productive and skilled to meet commercial and industrial demands (Suzana, Hamzah & Udin, 2011). The important role of TVET in Malaysia made the Ministry of Education come up with a transformation plan to improve the TVET system, specifically the program structure and teachers' recruitment. Among the two important transitions undertaken are the introduction of a four-year vocational college system to replace the previous two-year vocational school system and the recruitment of Malaysian Vocational Colleges teachers through the interim program. The recruitment of interim teachers aims to increase the number of Malaysian Vocational Colleges teachers with engineering or technical degree programs, who have links and experience with industries

Interim teachers in Malaysian Vocational Colleges are expected to be the first agents in providing technical knowledge and the latest hands-on experience in a real industry setting to students, so that they are better informed and prepared to enter the job market. However, since the interim teachers are new and come directly from industry without attending formal teacher training programs, there have concerns regarding their capability, especially their performance inside and outside the classroom, their social relationships with other teachers and with administrators. Several studies have reported on the issue of interim teachers' pedagogical competency. For instance, Goh, Qismullah and Wong (2017) identified a few problems in interim teachers' pedagogical competency, especially regarding their abilities to control the classroom and students' behavior, their teaching preparation, as well as their lack of pedagogical knowledge.

In some cases, some interim teachers have been in a comfortable working zone in their previous profession. Hence, the transition to becoming a professional educator is difficult because they cannot adapt to the new tasks and have difficulties in developing a positive relationship with their mentor teachers (Mohammad Hussain, 2016). Additionally, some of the interim teachers assume that they do not need any training for teaching as they are more skilled in the practical and technical area (Mohammad Hussain, 2016).

Training is required when there is a gap between the skills, knowledge and attitude of teachers with what is expected of them. The gap can be narrowed through in-service training to improve knowledge, skills or behavior of teachers in relation to their work (Sparks & Loucks-Horsley, 1989). In regard to interim teachers in Malaysian Vocational Colleges, they need specific in-service training because their qualifications and backgrounds are different compared to new teachers from educational degree routes. However, since the recruitment of interim teachers in Malaysian Vocational Colleges just began in 2016, there is no research examining the challenges faced and the training needs of interim teachers for a better transition to becoming professional educators. Hence, this study was conducted to identify the challenges faced by interim teachers in Malaysian Vocational Colleges, as well as to determine the support and training required by them.

2 CONCEPTUAL FRAMEWORK

This study utilized Charlotte Danielson's framework for teaching. This framework is one of the frameworks or models developed to understand the factors that influence and improve teachers' professionalism, which was introduced by Charlotte Danielson in 2007. The framework for teaching is also a road map that can meet the needs of new teachers or strengthen the teaching performance of veteran teachers (Danielson, 2007). Hence, it is also suitable for application in this study to enhance the professional development of interim teachers toward becoming better educators. Danielson's framework of teaching contains four main domains, which are the competencies that need to be mastered by teachers: 1) planning and preparation; 2) the classroom environment; 3) instruction; and 4) professional responsibilities (see Figure 1). In this study, all four competencies listed in Danielson's framework for teaching will be a basis used to identify the challenges faced by interim teachers in Malaysian Vocational Colleges, as well to determine the area of support and training sought by Malaysian Vocational Colleges' interim teachers.

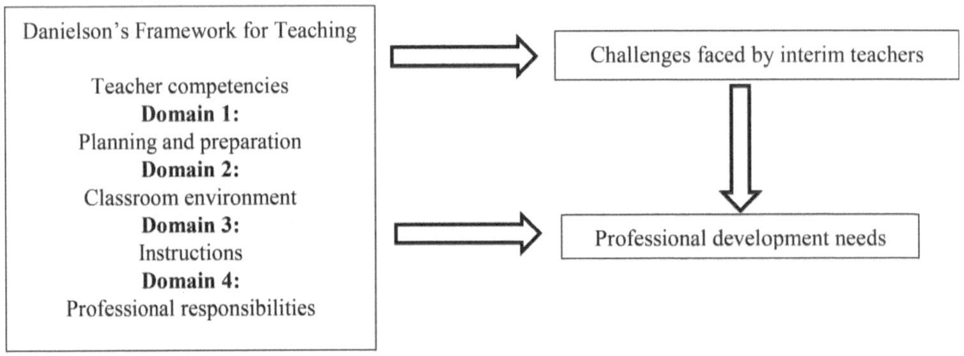

Figure 1. Conceptual framework (modified from Danielson's framework for teaching, 2007).

3 RESEARCH METHODOLOGY

This study used a qualitative method because the researchers were keen to find out rich and deep information regarding the challenges faced by interim teachers in Malaysian Vocational Colleges and their needs for professional development. Denzin and Lincoln (2011) state that a qualitative study allows the researcher to understand things in the informants' natural environment and helps the researcher to correctly understand or interpret the meaning of information brought by the informants. The data in this study were collected using a semi-structured interview protocol that allowed flexibility in collecting information and feedback from the informants. For example, the questions can be modified and arranged in a particular order and the level of language in the questions can be modified according to the needs in the interview process (Piaw, 2006). The interviews were guided by open-ended questions on issues related to motivation, job experiences and job satisfaction as interim teachers. For the purpose of this article, the main question was: 'How do you generally feel about being an interim teacher in Malaysian Vocational Colleges?' and 'What kind of in-service support do you seek in this profession'. Before the interviews were conducted, the supervisory committee members reviewed the interview questions regarding language, wording and relevance in order to assure validity.

The data collection process was conducted at two vocational colleges in Perak. The informants involved in this study consisted of five interim teachers. All of the informants were in the same teaching area, which is electrical technology. Since the number of respondents interviewed was minimal, the qualitative design helps an issue to be explored in-depth and in detail because this design generally generates information that is detailed, open and deep (Patton, 2002). The data from the interviews were transcribed and later coded and analyzed thematically to determine different and repeated ideas using Miles and Huberman's interaction model (Miles & Huberman, 1994).

4 RESEARCH FINDINGS AND DISCUSSIONS

The findings of this study indicate that the two biggest challenges faced by interim teachers in Malaysian Vocational Colleges are work stress, the difficulties in managing the classroom and students' behavior. They also seek more training and guidance from their mentors and senior colleagues regarding these issues.

4.1 *Work stress*

Copper et al. (1987) define work stress as an individual's inability to respond to the stimuli in the self and work environment, thereby having a certain impact to him or herself. The definition of work pressure in a school setting refers to the act of pressing, pushing, coercion, pressing conditions or stressful situations for teachers while working in schools. In school settings, teachers normally face the biggest challenge, which is to prepare themselves for the administrative workload given by school administrators. This burden is not just felt by trained teachers, but interim teachers also experience the same thing to the extent that some of the interim or new teachers have to delay the work given by their mentors or administrators. Due to the overloaded workload, they sometimes have to postpone the teaching and learning activities in order to manage other matters such as to prepare the quality assurance audit document needed by the Malaysian Qualification Agency (MQA) For example, one of the informants (GI 2 82) in this study mentioned that:

> At first, we thought it was just teaching. But in reality, we do other things. For instance, we are in class but suddenly we were called for a meeting. So, that is considered disrupting our teaching process because we had set the time today to do two hours of practical activity, so when there's a meeting we have to cancel the practical session. In MQA matters, we have to prepare many document that takes a lot of our time [GI 2 82].

The opinion of GI 2 82 is supported by another interim teacher, GI 3 49:

> Honestly, many documentation work needs to be done. Not just permanent teachers do it, interim teachers do it too. Sometimes we work, we haven't finished 'A', we have another small 'a'. We haven't even finished small 'a', there is 'B' already (chain of ad hoc workload). With the overloaded workload it is stressful and we lose focus on teaching [GI 3 49].

This study supports the work of Mohd Aderi et al. (2013), which found that the heavy workload and time constraints of new teachers contributes to the detriment of their teaching quality. In response to this issue, when the informants were questioned regarding the relevant professional development that they think is needed to help them overcome the work stress, surprisingly only one of them requested in-service training in stress management, while two of them requested training in time management. However, all of them mentioned that they need more support regarding documentation, especially in relation to the MQA audit.

4.2 *Student communication and behavioral management*

The Organization for Economic Co-operation and Development (OECD) reported that one-half of teachers in Brazil, Malaysia and Singapore spend 15% of teaching and learning time controlling their classes, whereby teachers lose time for lessons in order to manage students with problematic behavior in the classroom (OECD, 2014). Beginning or less experienced teachers normally believe that rules and discipline are important to manage their classroom. However, lack of experience made interim teachers have difficulty implementing rules and discipline in their classroom, especially in dealing with adolescent students' behavior. This is different from senior teachers, where their experience gave them confidence in effectively managing their classroom and students' discipline. The findings of the study show that the majority of interim teachers have problems in managing students' behavior or attitudes during teaching and learning activities. Some of the comments are:

> The students are a bit difficult to control… especially male students. When we teach in front, some of them even play with their mobile phones at the back, especially in theory classes [GI 2 64].

> Like when we first entered, we didn't know what to do if the students are wrong, we were afraid of the students [GI 2 66].

> It's worse in theory classes, some of students did not focus when we teach, some of them even sleep … [GI 3 53].

According to the informants, they seek more training and guidance from experienced mentors in dealing with classroom management and the student discipline issue. They need something that can help them to develop their confidence in managing their classroom and students or be able to communicate well with their students. One of the respondents mentioned that:

> It will be much helpful if I can sit together with my mentor so they can share any tips to develop good relationship with student [GI 2 66].

Regarding this issue, usually, classes that are not well-managed contribute to student disciplinary problems and this can prevent an effective teaching approach from occurring. Therefore, teachers' knowledge of the appropriate controlling strategy to be applied to students with behavior problems is the most important aspect requiring due attention. As a suggestion, the interim teachers need more training to be more sensitive to the environment as well as be able to identify the diversity of students in the classroom. Teachers need to build students' confidence in the teaching process, so that they will be confident to engage themselves voluntarily in the learning activities. Hence, teachers need to be trained so they can effectively hear students' opinions, as well as involve them in classroom activities. This can encourage students to speak up and promotes interaction among them (Arbaa, Jamil & Ahmad, 2017). The trust of teachers to provide learner autonomy in the process of teaching and learning will increase students' motivation to learn and cooperate.

5 CONCLUSION

This study looked at the aspects of challenges faced by interim teachers in vocational colleges, and the professional development needs of this group of teachers. Based on the findings, the interim teachers are looking for professional development programs that could help them in facing work stress and in managing students' behavior. Some of the professional development programs that have been suggested by the interim teachers are stress management, time management, managing students' behavior and classroom management. The professional development or in-service training provided to interim teachers in Malaysian Vocational Colleges will be a positive effort to enhance the abilities of the interim teachers in becoming professional educators. It is hoped that this study becomes a guideline to vocational college administrators, the Malaysian Institute of Teacher Education (IPGs) and the Technical and Vocational Education Division (BPTV) in identifying the professional developments required by interim teachers, in order to pave a better transition of the Malaysian Vocational Colleges interim teachers to become professional educators.

ACKNOWLEDGMENTS

This research study was supported by the 2017 Universiti Pendidikan Sultan Idris Research Grant. The title of this study is The Need for Professional Development Among Interim Teachers in Vocational Colleges (2017-0117-106-01).

REFERENCES

Arbaa, R., Jamil, H., & Ahmad, M. Z. (2017). Model bersepadu penerapan kemahiran abad ke-21 dalam pengajaran dan pembelajaran [The application of Integrated model of 21st century skills in Teaching and Learning]. *Malaysian Journal of Education (0126-6020), 42* (1).

Copper, C. L., Sloan, S. J. & Williams, S. (1987). *Occupational stress indicator management guide.* Oxford, UK: NFER-Nelson Publishing.

Danielson, C. (2007). *Enhancing professional practice: A framework for teaching* (2nd ed.). Alexandria, VA: Association for Supervision and Curriculum Development.

Denzin, N. K. & Lincoln, Y.S. (2011). *The SAGE handbook of qualitative research* (4th ed.). Thousand Oaks, CA: Sage Publications.

Goh, P.S.C., Qismullah, Y. & Wong, K.T. (2017). Lived experience: Perception of competency of novice teachers. *International Journal of Instruction, 10*(1), 21–36.

Miles, M.B. & Huberman, A.M. (1994). *Qualitative data analysis* (2nd ed.). Thousand Oaks, CA: Sage Publications.

Mohammad Hussain, M. A. (2016). *Novice career and technical education teachers' participation in professional development in the United States* (Doctoral dissertation, Pennsylvania State University).

Mohd Aderi, C. N., Mohd Sofian, M. L. & Asmawati, S. (2013). Teaching competency among Islamic educators in Malaysia. *World Applied Sciences Journal, 24*(2), 267–269.

OECD (2014). *A teachers' guide to TALIS 2013: Teaching and learning international survey, TALIS.* OECD Publishing. http://dx.doi.org/10.1787/9789264216075-en

Patton, M.Q. (2002). *Qualitative evaluation and research methods* (2nd ed.). Newbury Park, CA: Sage Publications.

Piaw, C.Y. (2006). *Kaedah dan statistik penyelidikan [Research Method and Statistic].* Kuala Lumpur, Malaysia: McGraw-Hill.

Siti Maspah, H. & Nor Azizah, M. S. (1995). *Pendidikan vokasional - Formal dan non-formal ke arah wawasan 2020 [Vocational education – Formal and non-formal toward vision 2020].* Paper presented at the Seminar Kebangsaan Pendidikan Negara Abad Ke 21 di UKM, Bangi.

Sparks, D., & Loucks-Horsley, S. (1989). Five models of staff development. *Journal of staff development, 10*(4), 40-57.

Suzana, N., Hamzah, R. & Udin, A. (2011). Professionalisme guru PTV dalam membentuk insan kendiri [Professionalism of PTV teachers in developing human capital]. *Journal of Edupress, 1*(2), 230–237.

TVET Towards Industrial Revolution 4.0– Hazirah Noh@Seth et al. (eds)
© 2020 Taylor & Francis Group, London, ISBN 978-0-367-24273-2

Work skills factor for mechanical engineering students in vocational high school

S. Hartanto, S.L. Ratnasari & Z. Arifin
Universitas Riau Kepulauan, Batam, Indonesia

ABSTRACT: Vocational education graduates are indicated as having a very low competence and cannot meet the expectations of the work requirement; it has an impact on a lower absorption of employment for vocational education regionally and nationally. In order to meet the needs of the job competence, vocational students should have good work skills. The purpose of this study was to determine the need for work skills consisting of soft skills and hard skills for vocational education students of mechanical engineering. This research was quantitative descriptive analysis conducted by using the Developing a Curriculum (DACUM) approach. The sample of the study consisted of 100 respondents, comprising of industry practitioners, vocational education practitioners, and relevant expert of vocational education in engineering. Based on the analysis, there are 27 items of soft skills and 67 items of hard skills recommended for employments for vocational students. Based on the analysis, the findings will be used as a reference for developing a lean-based learning model to improve the work skills of vocational students of mechanical engineering.

1 INTRODUCTION

Vocational education is organized as an effort to prepare individuals in reaching expected competency skill levels, in order to sustain the life of the individual, the workplace, and develop a career in the future (Calhoun & Finch, 1982, p. 60; Law No. 20 article 15, 2003; Hartanto et al., 2017). Graduates of vocational high school should be regarded as graduates who are ready to work, intelligent, have a competitive advantage, and a comparative and strong character as a working professional; therefore, the mastering of hard skills and soft skills must be covered very well to provide excellent and better-quality graduates in facing the world competition of work. Hard skills must be balanced with soft skills, and soft skills have a tendency to be a decisive factor in the recruitment process (Galuh, 2013). Vocational education has guaranteed the availability of workforce; however, the level of employment absorption of graduates of vocational high schools is not balanced with the condition of the hard skills and soft skills possessed by employees graduated from vocational high schools. The absorption of employment nationally for vocational senior high schools (SMK) is 10.87%, lower than for high school graduates who are at 20.52%, for junior high schools at 18.16%, and even for elementary school levels that are far greater at 42.23%, which means there are a lot of unemployment and high employment absorption differences between graduates of the education levels. For example, one of the provinces in Indonesia, Riau Islands, has been carrying out development with a complete infrastructure in several areas: agriculture, marine, and industry. In accordance with the engine works, the number of workers absorbed in the field of industry of manufacturing and machining in the Riau Islands was 41.20% (BPS, 2017). The majority of the industry is located in Batam, but the uptake of its workforce is still very low. An important issue must be resolved. It should be of great potential to resolve such issues, especially for vocational schools in order to meet the needs of the industry, the world of work and the increasing employment absorption of the vocational education.

Learning at SMK has not approached aspects of hard skills and soft skills comprehensively and thoroughly, so that graduates do not have quite enough of the hard skills and soft skills that industry needs. Developing hard skills and soft skills directly integrated with the work process directly forms the experience of hard skills and soft skills in the learning process. SMK can improve the competency skills through appropriate learning strategies to fit the needs of work, improving knowledge, attitudes, skills, and values that are needed for the job (Palmer, 2007; Lubis, 2010; Hartanto et al., 2017). Students should be provided with the knowledge, skills, attitudes, and values necessary in a real working environment. Integrated learning with the world of work provides experience to students and builds bridges between school education and the professional working world (Prosser & Quigley, 1949; Hartanto et al., 2017). Learning that is integrated with the real working world provides a huge benefit in the acquisition of vocational competence. Integration with the world of work is a form of environmental role to change the competence of vocational students in achieving sustainable development (Blum, 2008; Sousa, 2011).

The learning system should be established in accordance with the needs of the community. This analysis is an important step to map the special needs of work skills in the mechanical engineering departments of SMKs, to be used as a reference in composing the learning. The aim of this research is to determine the extent of needs of soft skills and hard skills for SMK students, as an effort to develop the work competence skills of students majoring in mechanical engineering.

2 LITERATURE REVIEW

The dynamic changes in the industry needs to be observed, so that a learning process that is similar to the industry standard is necessary. Learning is a process toward change. Learning is the process of obtaining mastery, knowledge, and habits, which are obtained from the process of learning (Bahri, 1996). The knowledge in the form of facts, concepts, procedures, and principles of the students' characteristics can be obtained by involving interaction with the external environment that can change someone's behavior (Bahri, 1996; Gagne & Briggs, 1992; Rukun et al., 2015). Vocational education must have the principle of change in accordance with the needs of society and technology. Prosser and Quigley's (1949, p. 34) second proposition: "*Effective vocational training can only be given where the training jobs are carried on in the same way with the same operations, the same tools and the same machines as in the occupation itself*". Vocational education should meet these standards in order to produce a competent workforce. Vocational education as one vehicle to prepare students for the world of work must be future-oriented: "*Socialize people into attitudes appropriate for the world of work, orientate people to understand the world of employment and to prepare for the choices and transition they will have to make on entering it, prepare them with specific skills and knowledge to apply in a direct way after entering employment*" (Adrian, 2005, p. 443).

Appropriate learning involves integrating it with industry. The main principle in industry is to achieve efficiency and high productivity. It is implemented by way of applying lean manufacturing. Lean is a production system that claims to create learning organizations through continuous improvements (Liker & Meier, 2006; Ståhl et al., 2015). "*The five primary elements for lean manufacturing are (1) manufacturing flow, (2) organization, (3) process control, (4) metrics, and (5) logistics*" (Feld, 2001). Lean manufacturing is a concept and the principles used in the company and the production process to maximize the work to achieve the maximum benefit by applying the five principles with no separation: "*Lean manufacturing, is primarily focused on designing a robust production operation that is responsive, flexible, predictable, and consistent*" (Feld, 2001, p. 21). Lonnie (2010) states that "*it is called Lean because, in the end, the process, it can run: using less material, requiring less investment, using less inventory, consuming less space and, using fewer people*". In order to achieve high efficiency and productivity at work, vocational school graduates must have the skills of work in accordance with the needs of the workforce. Work skills, skills that needs to be owned by individuals in the process of work, consist of hard skills and soft skills. Hard skills in vocational education are one type of skill that must be possessed by vocational students to achieve competence: "*There are some skills that are specific to needs in industry and manufacturing, the skills need will be very important as a defense to face of the information development and the environment rapidly*" (Hartanto et al., 2017, p. 157). Hard skills are abilities related to something that can be learned in education, aimed at improving intellectual ability. Coates (2006, p. 1) argues that "*hard skills are technical or administrative procedures related to an organization's core of the business*". Mazoota (2015, p. 1) states that "*Hard skills are skills where the rules stay the same regardless of which company, circumstance or people you work with*". Soft skills are the abilities of each individual, which cannot be seen, but the soft skills are the ability that play a huge role in one's life, which strongly support someone's ability, career and job (Hartanto et al., 2017; Robles, 2012). "*Appropriate soft skills play an important role in a successful career as well as during social interactions in the society. In addition, reviews on these skills are highly sought after by employers recruiting fresh graduates*" (Majid & Liming, 2012, p. 1036; Hartanto et al., 2017, p. 156).

Soft skills are "*character traits, attitudes, and behaviors-rather than knowledge or technical aptitude. Soft skills are the intangible, nontechnical, personality-specific skills that determine one's strengths as a leader, facilitator, mediator, and negotiator*" (Hartanto et al., 2017, p. 157; Robles, 2012, p. 457). Coates (2006, p. 1) says that "*They have to do with how people relate to each other: communicating, listening, engaging in dialogue, giving feedback, cooperating as a team member, solving problems, contributing in conducted some first-rate training and was well meetings and resolving conflict*". Chaturvedi et al. (2011, p. 5) state that "*the impact which the soft skills training could give in enhancing the output of hard skills. Soft skills play a crucial role in making students employable as it enables them to be flexible, positive to change, handle ever-increasing expectations of employers and to stay globally competitive*". Rani and Mangala (2010, p. 4) propose that the structure of the future work in the industrial world will eliminate more non-skilled jobs, which will be replaced with jobs that require skills, a high expertise in the areas of reading, computation, communication, and problem-solving or reasoning skills.

The work skills needed by industry is found by using the Developing a Curriculum (DACUM) needs analysis approach. This approach is used in a variety of methods to determine the accuracy of the use of learning and competency mapping by practitioners or specialists who have done the work and activities in accordance with the areas of expertise (Hartanto et al., 2017; Norton, 2004). The results of the analysis will show the gap between teachings and learning in a vocational school, with the work processes and activities in the industry. The gap found will be a basis for the development to improve the learning process through the lean-based learning model.

3 RESEARCH METHOD

The type of this research was descriptive quantitative, through the DACUM approach. This is a method analysis job/occupation technique that is recognized by industry practitioners, educators, and consultants as an effective way to identify the duties, tasks, and related information required for a job/occupation. It also provides an excellent source of data for management decision-making, developing a training program, human resource development, career planning, needs assessment, test development, job redesign, performance evaluations, and quality control planning (Norton, 2004, p. 3; Hartanto et al., 2017, p. 157). The sample of this study was 100 respondents by using random sampling. It consisted of teachers of a mechanical engineering department, a practitioner of the machining industry, and an expert/ lecturer for the vocational high school in mechanical engineering. The research instrument used was a questionnaire that is based on the prepared blueprint in accordance with the needs of work skills of the vocational high school of mechanical engineering. The questionnaire used a Likert scale (Hartanto & Fordiana, 2018, p. 2). Validity and reliability of research instrument used expert judgment that consisted of seven experts.

4 RESULTS AND DISCUSSION

The results of needs analysis of work skills are categorized into two parts: soft skills and hard skills, for SMK students who are majoring mechanical engineering. For engineering needs of work skills in mechanical engineering of vocational high school, questionnaires were used which were filled out by respondents by using four categories of the scale options of Likert: Very Important, Important, Less Important, and Not Urgent. Response to work skills needed by using the questionnaire was determined through the respondents' level of achievement. Decision criteria of the level of achievement of response were assigning a minimum of 50% of the total number of the respondents' preferences through a questionnaire to the category selection Agree and Strongly Agree. Based on the level of achievement of the indicator, new recommendations were elected for the establishment of indicators on soft skills and hard skills of students of vocational education in mechanical engineering. The results of the analysis are used in an attempt to build a new learning model of Lean-based learning model design for vocational high school students in an effort to improve work skills and competencies. The new learning model will be built by using a research and development method in the next step. The following describes further about the respondents' level of achievement indicators and tables on recommended soft skills and hard skills.

4.1 *Soft Skills*

Needs analysis of soft skills was categorized into two parts, namely general soft skills and specific soft skills in the field of machining jobs (Hartanto et al., 2017).

Table 1. Respondent response general soft skills category.

No	General soft skills category of machining jobs	Questionnaire category	
		Strongly agree	Agree
1	Demonstrating a willingness to develop a career	68	32
2	Showing ethics of communication	80	20
3	Showing the relationship among individuals	76	24
4	Showing good cooperation	92	8
5	Showing a high work ethic	78	20
6	Showing the action to solve the problem	56	42
7	Maintaining a presence on time	70	30
8	Indicating high initiative	56	42
9	Demonstrating honesty	82	18
10	Obeying all the rules work	64	34
11	Showing a responsible attitude	84	16
12	Showing a good adaptation in working	50	46

Table 2. Respondent response special soft skills.

No	Special soft skills in machining jobs	Questionnaire category	
		Strongly agree	Agree
1	Complying with the work process in accordance with the plans and design drawings	68	26
2	Adhering to the quotas for production	56	38
3	Showing the attitude of loyalty to the company	66	30
4	Motivated for training and teaching work processes	58	42
5	Showing the planning and operations according to the specifications of products	46	50
6	Demonstrating warming up the engine	44	52

(*Continued*)

Table 2. (*Continued*)

No	Special soft skills in machining jobs	Questionnaire category	
		Strongly agree	Agree
7	Showing the readiness of operational equipment	62	36
8	Showing the check engine units	64	36
9	Setting the machine according to product specifications	74	26
10	Demonstrating material handling right	52	44
11	Demonstrating health and safety at work	80	18
12	Suggesting caution in operating the machinery	74	26
13	Indicating maintenance and engine maintenance	66	34
14	Demonstrating checking the work according to standards of quality	62	34
15	Demonstrating off the machine according to the procedure	78	20

4.2 *Hard Skills*

Table 3. Respondent response hard skills – manufacture drawing.

No	Hard skills in machining jobs – manufacture drawing	Questionnaire category	
		Strongly agree	Agree
1	Understanding and applying the rules engine drawing techniques and workmanship mark	54	46
2	Understanding and demonstrating basic concepts and the command functions of Computer Aided Design (CAD)	46	54
3	Understanding and presenting detailed picture making, with etiquette of machine components with CAD drawings in accordance with International Standards Organization (ISO)	42	56
4	Analyzing and demonstrating the manufacture of engine components detailed images (projected images, image pieces and giving the size, tolerance, adjusting, a sign of craftsmanship and surface roughness value) with 2D/3D CAD	48	50

Table 4. Respondent response hard skills – lathe machining.

No	Hard skills in machining jobs – lathe machining	Questionnaire category	
		Strongly agree	Agree
1	Understanding and identifying the parts of a lathe by their type and function	48	46
2	Understanding, analyzing and identifying the suitability of the cutting tool lathe machines	24	70
3	Implementing and presenting procedures of eccentric turning technique	22	78
4	Evaluating and determining the procedure for turning technique eccentric	34	60
5	Analyzing and determining the making workpiece by using a faceplate	20	74
6	Implementing and making the technical procedure of making the workpiece assemblies, using various ways	30	52

Table 5. Respondent response hard skills – milling machine.

No	Hard skills in machining jobs – milling machine	Questionnaire category Strongly agree	Agree
1	Understanding and identifying the parts of the milling machine based on their type and function	36	56
2	Analyzing and identifying the use of cutting tools on milling machines	28	64
3	Evaluating and using a milling machine cutting parameters for different types of work	22	62
4	Implementing and using the standard operating procedure of fraising technique in all forms of the workpiece	38	58

Table 6. Respondent response hard skills machining – grinding machine.

No	Hard skills in machining jobs – grinding machine	Questionnaire category Strongly agree	Agree
1	Understanding and identifying the grinding machines for various kinds of work	34	56
2	Implementing and operating engineering machining surface grinders for various types of work	14	68
3	Choose and use cutting parameters of grinding machines for various kinds of work	14	58
4	Evaluating and using grinding machining techniques on various types of work	10	78

Table 7. Respondent response of hard skills machining – NC/CNC and CAM machining.

No	Hard skills in machining jobs – NC/CNC and Computer Aided Manufacturing (CAM) machining	Questionnaire category Strongly agree	Agree
1	Understand and identify the parameters and parts on a lathe and CNC milling	28	68
2	Implementing and operating all the procedures at each work machining lathe and milling CNC	24	72
3	Evaluating and repairing the failure of the work of a lathe and milling CNC	32	62
4	Analyzing and demonstrating CAM 2D and 3D for the process milling, lathe facing and drilling	30	58
5	Evaluating the use of CAM program through the simulation process on all machining jobs	28	58

NC/CNC: Numerical Control/Computer Numerical Control
CAM: Computer Aided Manufacturing

Table 8. Respondent response hard skills machining – industrial mechanical engineering.

No	Hard skills in machining jobs – industrial mechanical engineering	Questionnaire category Strongly agree	Agree
1	Understanding the concept and adhere to the appropriate maintenance manual/surgery	48	46
2	Understanding the types, functions and demonstrating major maintenance tools, mechanical and electrical	30	68
3	Understanding and classifying the types of disorders of the mechanical components of industrial machinery	42	52

(*Continued*)

Table 8. *(Continued)*

No	Hard skills in machining jobs – industrial mechanical engineering	Questionnaire category	
		Strongly agree	Agree
4	Analyzing the damage and performing minor repairs of industrial machinery mechanical components	30	64
5	Implementing and performing maintenance procedures/mechanical repair industry engine (compressor, pump, and motor gasoline)	46	46
6	Analyzing and demonstrating preventive maintenance in the mechanical industry	32	68
7	Analyzing and showing the reactive maintenance (reactive maintenance) in industrial machinery	26	64
8	Implementing and demonstrating a final check of mechanical and electrical components in industrial machinery	22	76
9	Implementing and demonstrating management workshop manufacturing jobs	20	54

Table 9. Respondent response hard skills machining – pneumatic and hydraulic systems.

No	Hard skills in machining jobs – pneumatic and hydraulic systems	Questionnaire category	
		Strongly agree	Agree
1	Analyzing disruption, damage and demonstrated improvements to components of pneumatic/hydraulic machinery industry	26	74
2	Understanding and demonstrating various types and concepts of fluid in the system of pneumatic/hydraulic to mechanical industry	38	54

Table 10. Respondent response hard skills machining – industrial machinery electrical system.

No	Hard skills in machining jobs – industrial machinery electrical system	Questionnaire category	
		Strongly agree	Agree
1	Understanding the concept and demonstrating the working principles of electrical symbols and diagrams on a production machine	30	66
2	Analyzing and demonstrating the maintenance work/disruption in the electrical circuit system of machine tools/production	32	62
3	Implementing and practicing the principles of electro-pneumatic circuit maintenance and electro-hydraulic	20	76

Table 11. Respondent response hard skills machining – design and drawing machine.

No	Hard skills in machining jobs – design and drawing machine	Questionnaire category	
		Strongly agree	Agree
1	Implementing and demonstrating the rules of drawing on machine construction drawing work	44	52
2	Analyzing and showing the results of analysis on engineering construction drawing machines	24	70
3	Evaluating and designing construction drawing machines with various types of connections	20	70
4	Analyzing and demonstrating the rules sign workmanship and price roughness on the picture detail of engine components	18	68
5	Applying and implementing rules engine component tolerances in the figure	32	60
6	Evaluating the changes and modify the image of engine components and product assemblies	30	56

Table 12. Respondent response hard skills machining – production control.

No	Hard skills in machining jobs – production control	Questionnaire category	
		Strongly agree	Agree
1	Understanding the types of types of production and implementing procedures in the process flow of the manufacturing industry	28	70
2	Implementing and measuring the performance of a production system in the manufacturing industry	24	70
3	Analyzing and planning the location and the standard of production in the manufacturing industry	26	66
4	Analyzing and improve production results which were not effective and efficient (waste)	30	64
5	Implementing and continuing improvement (continuous improvement) in the management of production	62	34
6	Understanding and applying in time production system in the manufacturing industry	32	66
7	Presenting and analyzing the design of an optimal control of production costs	22	76
8	Analyzing and managing the work environment according to the concept of production planning	14	84
9	Analyzing and processing data production forecasting total demand	24	66
10	Implementing and demonstrating the operation process map for manufacturing production	24	70
11	Applying and implementing procedures production process from beginning to end of production (materials, time, capacity)	32	68

Table 13. Respondent response hard skills machining logistics management.

No	Hard skills in machining jobs – logistics management	Questionnaire category	
		Strongly agree	Agree
1	Implementing and understanding the concept, the basic procedure for warehouse control (in, out, quality)	34	64
2	Processing and analyzing inventory balance	30	58
3	Implementing and analyzing the dismantling, removal and structuring effective and efficient goods	32	62

Table 14. Respondent response hard skills machining – warehouse governance.

No	Hard skills in machining jobs – warehouse governance	Questionnaire category	
		Strongly agree	Agree
1	Understanding the classification and demonstrating the use of the equipment used in the warehouse (the main equipment, support)	22	74
2	Implementing and carrying out the process of care and maintenance equipment and supplies warehouse	38	60
3	Understanding and implementing procedures for distributing and structuring the concept of shortening the distance/channel of distribution of goods from producers to consumers	30	64
4	Understanding and carrying out the principle of distribution of goods based on the accuracy of the type and specifications of the products, the accuracy of the value of the product, the accuracy of the number of products, on time and delivery place	50	48
5	Applying the data recording of goods in warehouse equipment used along with the use of the information system of warehousing	26	72
6	Implementing and demonstrating the process of using the material handling warehousing information systems	34	64

Table 15. Respondent response hard skills machining – safety at work.

No	Hard skills in machining jobs – safety at work	Questionnaire category	
		Strongly agree	Agree
1	Understanding and applying the basics and work safety system	62	38
2	Understanding and implementing safety procedures in all processes of machining work and industrial work processes	78	22
3	Analyzing and managing resources hazards and potential hazards posed to the manufacturing of machining jobs	72	28
4	Applying and demonstrating tools safety standards according to the procedure in the manufacturing of machining jobs	62	38

5 CONCLUSION

Based on the research with needs analysis of work skills that have been done, there are 27 items of *soft skills* chosen by the respondent. Soft skills are divided into two parts: there are 12 indicators for general soft skills and 15 for special soft skills indicator. For hard skills, there are 67 items of indicators needed to support vocational competence in mechanical engineering. Thus, indicators of soft skills and hard skills have been chosen according to the needs of the industry or the world of work in the field of mechanical engineering, which will be components of vocational work skills. Then the work skills must be possessed and mastered by students majoring in mechanical engineering of SMK as one of the necessary competences. Based on the needs analysis of work skills, then it will be set as a resource to develop a lean-based learning model. This learning model will be developed systematically as learning process improvement efforts to improve the work skills of vocational students majoring in mechanical engineering of SMK.

ACKNOWLEDGMENT

This work was supported in part by Research, Technology, and Higher Education of the Republic of Indonesia, No. 3/E/KPT/2018.

REFERENCES

Adrian, F. (2005). *The psychology of behaviour at work: The individual in the organization* (2nd ed.). New York, NY: Psychology Press.

Bahri, S. (1996). *Strategi belajar mengajar*. Jakarta, Indonesia: Rineka Cipta.

Blum. N. (2008). *Environmental education in Costa Rica: Building a framework for sustainable development? International Journal of Educational Development*, *28*(3), 348–358. doi:10.1016/j.ijedudev.2007.05.008

BPS (Badan Pusat Statistik). (2017). Kependudukan dan industri kelautan. Retrieved from Bps.go.id.

Calhoun, C.C. & Finch, A.V. (1982). *Vocational education concepts and operations* (2nd ed.). Belmont, CA: Wadsworth.

Chaturvedi, A., Yadav, A. & Bajpai, S. (2011). Communicative approach to soft and hard skills. *International Journal of Business & Management Research*, *1*(1). Retrieved from http://www.vsrdjournals.com.

Coates, D.E. (2006). People skill training. Retrieved from http://www.2020insight.net/docs4/peopleskills.pdf.

Feld, W.M. (2001). *Lean manufacturing: Tools, techniques, and how to use them*. New York, NY: St Lucie Press.

Gagne, R.M. & Briggs, L.J. (1992). *Principles of instructional design* (4th ed.). New York, NY: Holt, Rinehart and Winston.

Galuh, S. (2013). Banyak kegagalan tes karena soft skill. Retrieved from http://careernews.id/issues/view/1784-banyak-kegagalan-tes-karena-soft-skill.

Hartanto, S. & Fordiana, R. (2018). Learning needs analysis of vocational high school's chemical subjects in mechanical engineering department. *International Journal of Engineering & Technology*, *7*(3.25), 656–658. Retrieved from. https://www.sciencepubco.com/index.php/ijet/article/view/17818

Hartanto, S., Lubis, S. & Rizal, F. (2017). Need and analysis of soft skills for students of the mechanical engineering department of vocational high school. *International Journal of GEOMATE, 12*(30), 156–159. doi:10.21660/2017.30.TVET017. https://www.geomatejournal.com/node/599.
Liker, J. & Meier, D. (2006). *The Toyota way field book: A practical guide for implementing Toyota's 4Ps.* New York, NY: McGraw-Hill.
Lonnie, W. (2010). *How to implement lean manufacturing*. New York, NY: McGraw-Hill.
Lubis, S. (2010). Concept and implementation of vocational pedagogy in TVET teacher education. In *Proceedings of the 1st UPI International Conference on Technical and Vocational Education and Training, Bandung, Indonesia, 10–11 November 2010*. Retrieved from http://fptk.upi.edu/tvet-conference.
Majid. S & Liming. Z et al (2012). Importance of Soft Skills for Education and Career Success. International Journal for Cross-Disciplinary Subjects in Education (IJCDSE), Special Issue Volume 2 Issue 2, P.1036. Retrieved from:http://infonomics-society.ie/wp-content/uploads/ijcdse/published-papers/special-issue-volume-2-2012/Importance-of-Soft-Skills-for-Education-and-Career-Success.pdf.
Mazoota, A.R. (2015). *Workplace soft skills vs. hard skills – Which are more important*? Retrieved from http://www.armazzotta.com/.
Norton, R.E. (2004). *The DACUM curriculum development process*. In *14th IVETA Conference* (pp. 1–9).
Palmer, R. (2007). *Skills for work? From skills development to decent livelihoods in Ghana's rural informal economy. International Journal of Educational Development, 27*(4), 397–420. doi:10.1016/j.ijedudev.2006.10.003
Prosser, C.A. & Quigley, T.H. (1949). *Vocational education in a democracy*. Chicago, IL: American Technical Society. Retrieved from http://www.morgancc.edu/docs/io/Glossary/Content/PROSSER.PDF.
Robles, M.M. (2012). *Executive perceptions of the top 10 soft skills needed in today's workplace.* doi:10.1177/1080569912460400.
Rogalski, S.A. (2006). *Vocational education and training in the chemical industry in Germany and the United Kingdom*. ILO.
Rukun, K., Huda, A., Hendriyani, Y. & Hartanto, S. (2015). *Designing interactive tutorial compact disc (CD) for computer network subject. Jurnal Teknologi, 77*(23), 21–26. doi:10.11113/jt.v77.6682. https://jurnalteknologi.utm.my/index.php/jurnalteknologi/article/view/6682.
Sousa, D.A. (2011). Commentary: Mind, brain, and education: The impact of educational neuroscience on the science of teaching. *Learning Landscapes, 5*(1), 37–43.
Ståhl, A.C.F., Gustavsson, M., Karlsson, N., Johansson, G. & Ekberg, K. (2015). *Lean production tools and decision latitude enable conditions for innovative learning in organizations: A multilevel analysis. Applied Ergonomics, 47*, 285–291. doi:10.1016/j.apergo.2014.10.013
Undang-Undang Republik Indonesia No. 20. (2003). Tentang Sistem Pendidikan Nasional. Jakarta, Indonesia: BP Citra Jaya.

TVET Towards Industrial Revolution 4.0– Hazirah Noh@Seth et al. (eds)
© 2020 Taylor & Francis Group, London, ISBN 978-0-367-24273-2

The important elements of successful startup companies as a guideline for education

Noerlina
Department of Information Systems, School of Information Systems, Bina Nusantara University, Jakarta, Indonesia

W. Rusdyputra
Department of Computer Science, School of Computer Science, Bina Nusantara University, Jakarta, Indonesia

Sasmoko
Department of Primary Teacher Education, Faculty of Humanities, Bina Nusantara University, Jakarta, Indonesia

T.N. Mursitama
Department of International Relations, Faculty of Humanities, Bina Nusantara University, Jakarta, Indonesia

N.H. Abd. Wahid
School of Education, Faculty of Social Sciences and Humanities, Universiti Teknologi Malaysia, Johor, Malaysia

ABSTRACT: Recently, many people have been looking to leave the daily grind for something to feed their passion, not only adults but also teenagers. It was increasingly easy to start a business and there were many schools and also universities that had the subject of entrepreneur in their curriculum. With this early comprehension, people didn't need a long time to become an entrepreneur and have success. Startups had a high rate of failure, but the minority of successes included companies that had become large and influential. One of the purposes of this literature review was to analyze the aspects of successful startups to support the government's plan to increase the economics and education of the country. This research used a systematic literature review method, with a search process step of reviewing various sources of databases using keywords related to topics of the research; the obtained data was then classified based on the inclusion and exclusion criteria. There were 75 papers identified with the topic of research, which were then refined into 35 papers to review. The research identifies nine aspects of successful startup companies that are influential in determining their success, which are technology, funding, time, team, operation, market, experience, capital efficiency, and other external factors. More validation will be needed to verify that all aspects can be applied to all types of company and product.

1 INTRODUCTION

Innovation and startup can help to drive a nation's economy and education forward as a key engine of economic growth, with an understanding of the concepts of entrepreneurship from an early stage that every school and university has already applied. Thanks to the lean startup process and technology advances, entrepreneurs are scaling companies to sizable revenues with smaller teams and less cash compared with a few years ago. The cycle of innovation is speeding up and talented entrepreneurs are ready to take over to invent the next disruptive technologies [11]. In the world of business, the word 'startup' goes beyond a company that is just getting off the ground. The term startup is also associated with a business that is typically technology-oriented and has high growth potential. A startup has some unique challenges, especially in regard to financing. That's because investors are looking for the highest potential return on investment, while balancing the associated risks [12].

The broad majority of successful businesses, corporations and entities were at one time a startup. As per the statistics, nine out of ten startups fail. It has been perpetually the necessity of entrepreneurs to grasp the key factors concerned in making a winning business. Every bourgeois needs his/her hypothesis to figure out what might result in a winning enterprise. They require a product that is likable to their customers and for which they can induce sufficient traction. Some factors that help in creating a successful enterprise are traction, capital, management, skilled individuals, a viable product, and so on [1]. Therefore, startup companies need to find solutions to overcome their problems.

2 THEORETICAL BACKGROUND

New companies that are still looking for a validated business model are startup companies. A new company is not necessarily a startup, which is about the business model and about the success of the new products and services offered in the market [39].

The entrepreneur is usually a sole proprietor, a partner, or the one who owns the majority of the shares in an incorporated venture. However, people who can do both have the most potential to become successful. Most experts agree that creating an innovative business is at least equal in difficulty to creating an innovative product or service. It requires a unique set of skills, personal attributes and a mindset that many people don't have or choose to learn. In fact, there are many who are not choosing to build and nurture their technical skills. Many people may be want to develop the business and personal attributes associated with entrepreneurs. Technical entrepreneurs often believe that great products will ultimately lead to a successful business and only good business skills can accelerate this process. In fact, there are trade-offs involved between a focus on the technical aspects and business aspects. Entrepreneurs are able to find the right balance —technical perfectionism will need to adjust, just like minimal focus on building the business will cause problems for entrepreneurs, even with a strong product (13).

One of the first decisions a startup has to make is the appropriate corporate structure. Creating a company as an independent entity can help founders to protect their personal assets from potential liability claims, protect intellectual property, and provide tax benefits (11).

3 RESEARCH METHOD

This research used a systematic literature review method to identify aspects of startup success.

3.1 *Search Process*

1) Sources: Science Direct (www.sciencedirect.com), IEEE Digital Library (ieeexplore.ieee.org), ACM Digital Library (dl.acm.org), Emerald Insight (www.emeraldinsight.com), Wiley Online Library (onlinelibrary.wiley.com), Springer (link.springer.com), Taylor and Francis (www.tandfonline.com).
2) Keywords: Startup success, successful startup, and method of success, startup, recipes of success, aspects of startup, elements of startup.
3) Search strings: Startup and success, startup and success, 'start-up business' and success, 'start-up success' and business, 'start-up success' and business, start-up success or start-up success and aspects, startup or startup and success.

3.2 *Inclusion and Exclusion Criteria*

In this phase, the data that was found from the search process was classified as including the criteria of this research, or else was excluded based on the exclusion criteria:

1) Inclusion criteria: research must be in the main topic area; research must be relevant to research questions; research that describes aspects of startup success; research that describes academic journals or conferences or books only; papers based on the systematic literature method, descriptive analysis, qualitative or quantitative analysis; papers consisting of author's name, institution and country of institution.
2) Exclusion criteria: papers that discuss the procedures that are used to build startup business; papers that only focus on the technical aspects of startups; papers that show duplicate reports of the same research; papers based on opinions, theses, panel discussions; paper redundancy.

3.3 *Data Extraction*

In this phase, the data that was found from the search process is extracted according to three steps of data collection, which are:

1. Research found: paper match with search process.
2. Candidate research: paper collected based on title and abstract.
3. Selected research: paper read carefully starting with the introduction, analysis results and conclusions.

Table 1. Data extraction.

Source	Found	Candidate	Selected
IEEE	10	7	4
ACM	14	7	6
Emerald	12	7	5
Wiley	12	9	7
Springer	10	8	5
Taylor and Francis	5	3	1
Science Direct	11	9	6
Others	1	1	1
Total	75	51	35

4 RESULTS AND DISCUSSION

The selected papers have been extracted according to the inclusion criteria; the next step of this method is the analysis of the results based on demographic trends and characteristics; then the findings and final results.

4.1 *Publishing Outlets*

As shown in Table 2, 35 research papers were identified on this topic, consisting of 22 journals, eight proceedings and five books.

4.2 *Author Academic Background*

As shown in Table 3, most of the authors' academic backgrounds for this research topic lie in business management, which accounts for 63%.

Table 2. List of study materials.

ID	Title	Source	Type	#
S1	Predicting....[14]	IEEE	Proceeding	4
S2	Bringing....[15]	IEEE	Proceeding	
S3	A Success....[16]	IEEE	Proceeding	
S4	Determinants....[17]	IEEE	Proceeding	
S5	Evaluating....[18]	ACM	Journal	2
S6	The Essential....[19]	ACM	Journal	
S7	Organize,....[20]	ACM	Proceeding	3
S8	On the Critical....[21]	ACM	Proceeding	
S9	Measuring....[22]	ACM	Proceeding	
S10	Pattern....[23]	ACM	Journal	1
S11	The Power....[24]	Emerald	Journal	5
S12	Startup....[25]	Emerald	Journal	
S13	For Local....[26]	Emerald	Journal	
S14	Internet....[27]	Emerald	Journal	
S15	The Ingredients....[28]	Emerald	Journal	
S16	First....[30]	Wiley	Journal	5
S17	Startup....[29]	Wiley	Journal	
S18	Determinants of....[31]	Wiley	Journal	
S19	Small....[32]	Wiley	Journal	
S20	Recipe....[33]	Wiley	Journal	
S21	Survival....[34]	Wiley	Book	2
S22	Success,....[35]	Wiley	Book	
S23	Startup [36]	Springer	Book	1
S24	How do....[37]	Springer	Journal	3
S25	Startups....[38]	Springer	Journal	
S26	Expert....[3]	Springer	Journal	
S27	Unfamiliar.....[1]	Springer	Proceeding	1
S28	Psychological....[2]	Taylor and Francis	Journal	1
S29	Questionnaire....[4]	ScienceDirect	Journal	1
S30	From....[5]	ScienceDirect	Book	1
S31	The Use....[6]	ScienceDirect	Journal	5
S32	Exploring....[7]	ScienceDirect	Journal	
S33	Why Startup....[8]	ScienceDirect	Journal	
S34	Entrepreneur....[9]	ScienceDirect	Journal	
S35	Startup .... [11]	KPMG	Book	
Total				35

Table 3. Authors' academic background.

Academic Background	#	%
Business Management	33	63
Computer Science	5	9
Economic	6	11
Engineering Science	3	6
Geographical Sciences	2	3
Information Systems	5	9
	53	

4.3 *University Affiliation According to Country*

As shown in Table 4, the most productive affiliation in this topic is the USA, generating 13 papers from 28 authors.

Table 4. University affiliation according to country.

Country	Papers	%	Authors	%
Brazil	1	3	1	1
China	1	3	3	4
England	7	20	13	19
France	1	3	1	1
Germany	2	6	6	9
India	1	3	3	4
Japan	1	3	1	1
Korea	2	6	4	6
Malaysia	1	3	1	1
Netherlands	1	3	2	3
Poland	1	3	1	1
Spain	1	3	1	1
Switzerland	1	3	4	6
Uganda	1	3	1	1
United States	13	37	28	40
Grand Total	**35**		**70**	

4.4 *Findings and Final Result*

All the selected research has identified aspects of startup success in nine aspects, as shown in Table 5.

Table 5. Final result.

Aspect	Description
Technology	Technology aspects that impact the startup such as application and database. Successfully made a technology transfer, within the framework of a key partnership, rather than a pay-as-you-go principle, and part of a trend toward virtualization in enterprise Information Technology (IT). New device platforms and new models for monetizing applications have attracted application developers to the mobile Internet.
Funding	Companies have to improve their chances of getting the appropriate financing to meet specific needs. Initial fuel for any startup, a seed fund helps startups through their initial hurdles and at the same time act as an accelerator too.
Time	Timing is a very important thing in startup success. If the idea comes too early and consumers aren't ready for it, they won't readily adopt the system. If the idea comes too late, a number of distinct rivals may already be in front of the target audience of the company. Time and money are the two important parts of any startup.
Team	Prior to its incorporation, the founders also benefited from advisory support from a solid serial entrepreneur ecosystem.
Operationalization	The operation is keeping the company running and 'keeping the trains running on time' by building lightweight processes and controls. Three potential success indicators: survival, size of turnover, and number of personnel employed. Persons who take risks in operating in the market and whose income is determined by the financial performance of their market activities.
Market	The best technology proposal for customers while being extremely adaptive and supportive to customers' custom design requests.
Experience	Having the experience in the industry your startup exists in is irreplaceable when it comes to making your startup succeed.
Capital efficiency	The company's policy had always been to limit external funding to the strict minimum, and ensure sufficient levels of investment in R&D.
Other external factors	The factors noted in the literature that are associated with the success of new firms such as inflation, government regulation, and others.

5 CONCLUSION

Based on Table 5, there are nine aspects of a successful startup company that are influential in determining its success, which are technology, funding, time, team, operations, market, experience, capital efficiency, and other external factors. This research will need validation testing to verify that all aspects can be applied to all types of company and product.

REFERENCES

1 Johnson, A. (2017). Unfamiliar brands and exaggerated warranty: is it a recipe for success? In *proceedings of the 2016 academy of marketing science (ams) annual conference*

2 Baluku, M.M. (2016). Psychological capital and the startup capital–entrepreneurial success relationship. *Journal of Small Business and Entrepreneurship, 28*(1), 27–54.

3 Chossat, V. (2003). Expert opinion and gastronomy: The recipe for success. *Journal of Cultural Economics, 27*(2), 127–141.

4 Staniewski, M.W. (2018). Questionnaire of entrepreneurial success — Report on the initial stage of method construction. *Journal of Business Research, 88*, 437–442.

5 Picken, J.C. (2017). From startup to scalable enterprise: Laying the foundation. *Business Horizons, 60*(5), 587–595.

6 Dalmarco, G. (2017). The use of knowledge management practices by Brazilian startup companies. *RAI Revista de Administração e Inovação, 14*(3), 226–234.

7 Olugbola, S.A. (2017). Exploring entrepreneurial readiness of youth and startup success components: Entrepreneurship training as a moderator. *Journal of Innovation & Knowledge, 2*(3), 155–171.

8 Blagburn, N. (2016). Why startup boards matter. *Technovation, 57-58*, 45–46.

9 Devece, C. (2015). Entrepreneurship during economic crisis: Success factors and paths to failure. *Journal of Business Research, 69*(11), 5366–5370.

10 JaeShin, I. (2014). Loss prevention at the startup stage in process safety management: From distributed cognition perspective with an accident case study. *Journal of Loss Prevention in the Process Industries, 27*, 99–113.

11 KPMG. (2013). *Startup success*. New York, NY: KPMG.

12 Landau, C. (n.d.). *What's the difference between a small business venture and a startup?* Bplans. Retrieved from https://articles.bplans.com/whats-difference-small-business-venture-startup/

13 Zwilling, M. (2014). *Startuppro: How to set up*.

14 Amar Krishna, A.A.A.C. (2016). Predicting the outcome of startups: Less failure, more success. In *Proceedings of the 2016 IEEE 16th International Conference on Data Mining Workshops, Barcelona, Spain*.

15 Abelé, N. (2016). Bringing a mems startup to success from lab to market, with a fabless business model. In *Proceedings of the 2016 IEEE 29th International Conference on Micro Electro Mechanical Systems, Shanghai, China*.

16 Hayashida, H. (2007). A success factor of a digital material startup company. In *PICMET '07-2007 Portland International Conference on Management of Engineering & Technology, Portland, OR* (pp. 1322–1322). doi:10.1109/PICMET.2007.4349454

17 Reddy, S., Jaju, A. & Kwak, H. (2000). *Determinants of internet startups success*. IEEE Engineering Management Society.

18 Cusumano, M.A. (2013). Evaluating a startup venture. *Technology Strategy and Management, 56* (10), 26–29.

19 Kalmanek, C. (2012). The essential elements of successful innovation. *ACM SIGCOMM Computer Communication Review, 42*(2), 105–109.

20 Lehner, F. & Fteimi, N. (2013). Organize, socialize, benefit – How social media applications impact enterprise success and performance. In *Proceedings of the 13th International Conference on Knowledge Management and Knowledge Technologies*.

21 Li, J. (2005). On the critical success factors for B2B e-marketplace. In *Proceedings of the 7th International Conference on Electronic Commerce, China*.

22 Lehner, F. (2014). Measuring success of enterprise social software: The case of hypervideo. In *Proceedings of the 14th International Conference on Knowledge Technologies and Data-Driven Business, Austria*.

23 Staff, U. (2004). *Pattern of success*. ACM.

24 Allen, T.J. (2016). The power of reciprocal knowledge sharing relationships for startup success. *Emerald Insight*, *23*(3), 636–651.
25 Al-Darayseh, M.M. (1993). *Startup businesses — Selecting the right form: Tax and limited liability factors*. Emerald Group Publishing.
26 Emerald Publishing. (2017). For local startup success, look abroad: The influence of international entrepreneurship on born globals. *Strategic Direction*, *33*(5), 36–38.
27 Finkelstein, S. (2001). Internet startups in the new economy: So why can't they win? *Journal of Business Strategy*, *22*(4), 16–21.
28 Moghaddam, K. (2015). The ingredients of a success recipe: An exploratory study of diaspora opportunity entrepreneurship. *South Asian Journal of Global Business Research*, *4*(2), 162–189.
29 Hahn, G. (2016). Startup financing with patent signaling under ambiguity. *Journal of Financial Studies*, *46*(1), 32–63.
30 Zhao, Y.L. (2014). First product success: A mediated moderating model of resources, founding team startup experience, and product-positioning strategy. *The Journal of Product Innovation Management*, *32*(3), 441–458.
31 Schutjens, V.A. (1999). Determinants of new firm success. *Regional Science*, *79*(2), 135–159.
32 Merrett, C.D. (2004). Small business ownership in Illinois: The effect of gender and location on entrepreneurial success. *The Professional Geographer*, *52*(3), 425–436.
33 Coomber, S. (2014). Recipe for success. *Business Strategy Review*, *25*(2), 54–54.
34 Owen, J. (2008). *Survival and success*. Chichester, UK: Wiley.
35 Raspin, P.G. (2008). *Success, strategy and understanding*. Chichester, UK: Wiley.
36 Ready, K. (2011). *Startup*. New York, NY: Springer.
37 Frederiksen, D.L. (2017). How do entrepreneurs think they create value? A scientific reflection of Eric Ries' Lean Startup approach. *International Entrepreneurship and Management Journal*, *13*(1), 169–189.
38 McIlroy, Y. (2017). Startups within the U.S. book publishing industry. *Publishing Research Quarterly*, *33*(1), 1–9.
39 Blank, S.G. & Dorf, B. (2012). *The startup owner's manual: The step-by-step guide for building a great company*. Pescadero, CA: K&S Ranch.

TVET Towards Industrial Revolution 4.0– Hazirah Noh@Seth et al. (eds)
 ISBN 978-0-367-24273-2

Socio-economic determinants of students' academic achievement in building technology

M.S. Nordin & K. Subari
Universiti Teknologi Malaysia, Johor Bahru, Johor, Malaysia

Y.I. Salihu
Federal College of Education (Technical),Bichi, Kano, Nigeria

ABSTRACT: Over the years, the achievement of students in building technology has continued to decrease and this has been an issue of concern to major stake holders in vocational and technical education in north-central of Nigeria. There are some factors which influenced the academic performance of students such as socioeconomic factor. Therefore, this study investigated socio-economic determinants of students' academic achievement in building technology in North-central Nigeria. Data for this survey research design which collected from 113 students were analysed using descriptive and inferential statistics. The results of the study showed that building technology education in Nigeria is male dominated and majority of colleges of education students in North-central Nigeria are under 20 years. Socio-economic determinants of students' academic achievement in building technology include gender, age, study hours, family size, stipends per month, group reading, students' health status, credit pass in SSCE and truancy. The challenges facing building technology students in the study area include epileptic power supply, inadequate ICTs, lack of functional building technology workshop among others.

1 INTRODUCTION

Vocational and technical education has continued to receive global attention because of its strategic position in national economic growth and development. Okoye & Arimonu (2016) described technical and vocational education as a comprehensive term in the educational process involving, in addition to general education, the study of technologies and related sciences and acquisition of practical skills, attitudes, understanding and knowledge relating to occupations in various sectors of economic and social life. Vocational and technical education equips the recipients with practical and applied skills as well as basic scientific knowledge. According to Kehinde & Adewuyi (2015), vocational and technical education is fundamental to the development and industrialization of nations. This is because, the saleable skills and competencies that are needed for economic growth and development of the nation are embedded in vocational and technical education. In affirmation, Manyindo (2005) stated that vocational and technical education when properly developed provides the youths with knowledge, skills and training that satisfy the human resource demand of the nation.

National policy on education of Federal Republic of Nigeria (2014) enumerated the objectives of vocational and technical education in Nigeria to include: (i) provision of trained manpower in applied science, technology and commerce particularly at sub-professional grades, (ii) provision of technical knowledge and vocational skills necessary for agricultural, industrial, commerce and economic development, (iii) provision of people who can apply specific knowledge to the improvement and solution of environmental problems for the use and convenience of man, (iv) giving an introduction of professional studies in engineering and other technologies, (v) giving training and impact the skill leading to the production of craftsmen, technicians and other skilled personnel who will be enterprising and self-reliant and (vi) enabling our young men and women to have an intelligent understanding of the increasing complexity of technology. Aspects of vocational and technical education in Nigeria education system include: agriculture, home economics, business education, electrical and electronics technology, metal work technology, mechanical or automobile technology, woodwork technology and building technology. The focus of this study is on building technology.

Building technology according to Maurice et al. (2017) is one of the vocational and technical education courses that focus on the application of engineering principles and technology to design and construction of buildings. As described by Okoro (2010), building technology involves the technical methods, skills, processes, techniques, tools and raw materials needed for the construction and maintenance of buildings. Building technology borders on how technical and vocational education can be used to improve the quality of life of the individual and society by equipping learners with relevant scientific and technological skills, attitudes and values in building technology for livelihood. While justifying the significance of building technology, Egboh (2009) noted that the major distinction between an advanced country and a developing one is to a large extent the difference between their levels of scientific and technological development in various areas of technology education, building technology inclusive. The objectives of building technology at college of education or ordinary national diploma levels in Nigeria as reported by UNESCO (2001) include the provision of technicians that can assist the professional builder in the areas of production of simple buildings, maintenance of simple buildings, management of small projects, costing of simple construction works, cost control techniques in minor construction and engineering works and selection of materials and techniques for new building systems. Unfortunately, these objectives are far from been achieved with the current poor interest and achievement of students in building technology most especially at colleges of education level. Kolawole & Dele (2002) noted that students' academic performance is one of the current educational problems of public interest based on current poor level of student's academic performance especially in various higher institutions. Over the years, the achievement of students in building technology has continued to decrease and this has been an issue of concern to major stake holders in vocational and technical education in Nigeria and north-central Nigeria in particular.

Poor academic achievement, according to Aremu & Sokan (2003) is a performance that is adjudged by the examinee or testee as falling below an expected standard. Igberadja (2016) acknowledged the worrisome level of poor academic achievement of students in vocational technical education subjects in Nigeria. May et al. (2007) observed that technical and vocational education are very much still neglected which consequently is robbing the country of the economic development to be contributed by graduates of technical and vocational education. Similarly, Diraso et al. (2013) reported that students' prospects for career pursuits in engineering and technical and vocational education were being hampered by their poor performances in core engineering and vocational technical education subjects. In quest for solution the problem of poor achievement of students in vocational and technical education, Igberadja (2016) noted that various factors related to teachers' qualification, age, experience, and gender may affect the academic performance of students in science and technical education. The report of National Board for Technical Education (2012) showed that factors that worsen poor achievement of students in vocational and technical education towards an effective career choice are poor background of students, truancy, and language of instruction of the teacher, among others. In addition, Unongo (2015) expressed concern over the growing gap between the industrial skill needs and the acquired skill level of students in vocational and technical education. Hence, it is crystal clear that academic achievement of students in Building technology is greatly influenced by an interplay of some factors and socioeconomic variables. It is based on this background that this study investigated socio-economic determinants of students' academic achievement in Building Technology in North-central Nigeria. Specifically, the study identified challenges facing students of Building technology in the study area and also socio-economic determinants influencing students' achievement in Building technology.

2 METHODS

The study was carried out in North-central geopolitical zone of Nigeria using survey research design. North-central Nigeria is made up of 6 states which include Kogi, Niger, Benue, Kwara, Plateau, Nassarawa and the Federal Capital Territory.

2.1 *Sampling*

Random sampling technique was used to select 113 NCE II students of Building Technology drawn from the four Colleges of Education.

2.2 *Instrument*

Data for this study were obtained from both primary and secondary sources. Through the use of a structured questionnaire, data on socioeconomic attributes of the students such as age, gender, study hours, parents' education background, family size, stipends received per month and students' health status among others were collected. To avoid deception on academic achievement, data on each of the students' Cumulative Grade Point Average (CGPA) were collected from departmental academic records.

2.3 *Analytical Techniques*

Data collected were analysed using descriptive statistics and ordinary least square (OLS) multiple regression analysis. Descriptive statistics such as frequency, percentages, mean and standard deviation were used to identify the challenges facing students of Building technology in Colleges of Education in the study area. The intensity of the challenges facing the students was graded into 4-point rating scale of Very Serious, (VS) = 4, Serious (S) =3, Less Serious (LS) =2 and Not Serious (NS) = 1. The mean ratings of the students based on the 4-point rating scale were interpreted using boundary limit as stated in Table 1 below:

Therefore, items with mean values between 3.50 – 4.00 were interpreted as very serious challenges, those between 2.50 – 3.49 were interpreted as serious challenges, and those between 1.50 – 2.49 and 1.00 – 1.49 were interpreted as less serious and not serious challenges respectively.

Ordinary Least Squire multiple regression analysis was employed to estimate determinants of academic achievement of NCE II students of Building Technology. Students' CGPA was used as a proxy of their academic achievement. The implicit form of the regression model is stated thus as follow in (1):

$$Y = f(X1, X2, X3, X4, X5, X6, X7, X8, X9, X10, X11, X12, X13, X14 + \ell) \tag{1}$$

where Y = Students' Cumulative Grade Point Average (CGPA); X1 = Gender of the students (Male = 1; Female = 0); X2 = Age of the students (in number of years); X3 = Number of study hours (in number); X4 = Parents' education background (Literate = 1, Illiterate = 0); X5 = Family size (in number of persons); X6 = Family type (Nuclear = 1, Polygamous = 0); X7 = Stipend per month (in naira ₦); X8 = Engagement in group reading (Yes = 1, No = 0); X9 =

Table 1. The interpretation of 4-point rating scale

Response Categories	Ordinal values	Boundary limits
Very Serious (VS)	4	3.50 – 4.00
Serious (S)	3	2.50 – 3.49
Less Serious (LS)	2	1.50 – 2.49
Not Serious (NS)	1	1.00 – 1.49

Frequency of parents' visits to school (in number of times); X10 = Students health status (Having health challenge = 1, no challenge = 0); X11 = Number of credit pass in SSCE/NECO (in number); X12 = Truancy (number of time absent from lecture in last semester); X13 = Accommodation status in School (Campus = 1, Off-campus = 0); and X14 = Involvement in Social Activities (Yes = 1, No = 0)

The explicit form of the linear model is as follows in (2)

$$Yc = b0 + b1x1 + b2x2 + b3x3 + b4x4 + b5x5 + \dots \dots + e \tag{2}$$

Three functional forms: linear, semi-log and double-log were estimated using the Ordinary Least Square. This was considered necessary in order to select the functional form with the best fit. In the semi-log and double log forms, 0 values were not logged because, the number 0 is undefined for log.

3 RESULTS AND DISCUSSION

The pie chart in Figure 1 below showed that majority of 80.5% of the NCE II students of Building Technology are males while the remaining 19.5% of the students are females. This indicates that Building technology education in Nigeria is male dominated. The findings of this study agreed with that of Egun & Tibi (2010) who identified wide gender gap of students' enrolment in vocational education in favor of males. Similarly, the findings of this study also conformed to that of Dokubo & Deebom (2017) whose findings showed gender disparity in students enrollment in technical education with very low enrolment of females in technical education in Nigeria.

The bar chart in Figure 2 revealed that majority of 43.4% of the NCE II students of Building Technology are within 15 – 20 years age bracket, 31.8% are within 21 – 25 years age bracket, 17.7% of the students are within 26 – 30 years age bracket while 7.1% of the students are 31 years and above. The result shows that most of NCE II students in North-central Nigeria are very young with few old students. In affirmation of this result, Ademola et al. (2014) found that about sixty-three percent of the population of undergraduate students in Nigeria tertiary institution are under the age of twenty years.

3.1 *Socio-economic Determinants of Academic Achievement of Building Technology Students*

Table 2 shows the results of the regression analysis on determinants of students' academic achievement in Building technology in terms of CGPA. Three functional forms (linear, semi-log and double-log) were statistically estimated, the double-log functional form had the best fit, based on the R^2 value of (0.71), number and levels of significance of independent variables and signs. For instance, the R^2 value of 0.71 indicates that the significant variables in the model are responsible for about 71% of the variation in achievement (CGPA) of the Building Technology students. The F-value of (27.854) implies that the overall equation was significant

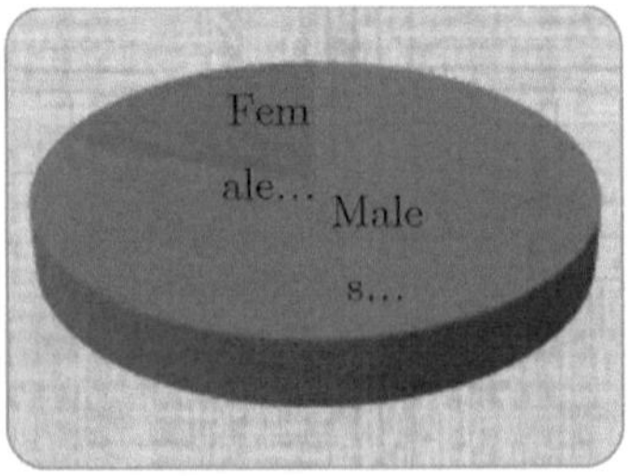

Figure 1. Distribution of students by gender.

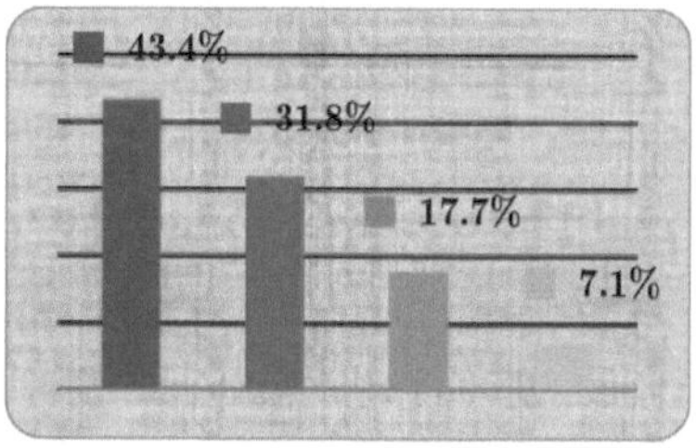

Figure 2. Distribution of students by age.

at ($p<0.01$) while Durbin-Watson (DW) of 2.230 indicates absence of autocorrelation. Out of the 14 independent variables specified in the model, nine significantly influence achievement (CGPA) of NCE II students of Building Technology. The significant variables include: gender, age, study hours, family size, stipends per month (₦), group reading, students health status, number of credit pass in SSCE/NECO and truancy.

The coefficient gender was significant at 5% and positively related with academic achievement of NCE students of Building technology. The result implies that male students have higher academic achievement (CGPA) than their female counterparts. Igberadja (2016) found that various factors related to qualification, age, experience, and gender may affect the academic performance of students in science and technical education. Age of the students was highly significant at 1% but negatively related with academic achievement in Building technology. This indicates that the higher the age of the students, the lower their academic achievement. Hence, younger students in the study have higher CGPA than the older ones. An increase in age of individual may translate to increase in responsibilities that may negatively interfere with academic programme of students. The findings of this study agreed with that of Ogunsola et al. (2014) who found that several factors such as age, gender, geographical belonging, ethnicity, socioeconomic status, language, religious affiliations, family size and type play significant roles in determining the academic achievement of students. The coefficient of study hour was highly significant and positively related with CGPA at 1%. This result indicates that the more the number of study hours of students increase, the more their achievement in terms of higher CGPA. This conformed with *a priori* expectation. The findings of this study conformed with that of Akinoso (2017) who found that students' ability to study at home and student attitude to instruction significantly predicted their academic achievement in Mathematics. Alufohai (2016) found that effective instruction and text messaging significantly influence student's academic achievement.

The coefficient of family size was significant at 5% but negatively related with academic achievement in Building technology. This showed that the higher the number of the people in students family, the lower their CGPA. The implication of the inverse relationship of family size with achievement could have resulted from inadequate finance for necessary instructional materials as a result of large family size the student come from. In this case, students from small sized families have better academic achievement. Stipend (pocket money) received from parent and guidance was highly significant at 1% and positively related with academic achievement in Building technology students. This suggests that the higher the higher the money made available to students, the more their ability to purchase the necessary instructional materials and faculties for quality learning, hence, the higher their CGPA. Abdullahi et al. (2015) in a study equally found that family size and feeding, provision of resource materials, visits to schools, provision of pocket money, and residential type (accommodation) statistically and significantly influence academic achievement of students in schools. Jeynes (2002) reported that provision and access to various educational materials and resources influence academic achievement of students. The coefficient of group reading of Building technology students was highly significant at 1% and positively related with CGPA. This was expected and students in reading groups share ideas and collaborate with one another for increased academic achievement in terms of higher CGPA.

Table 2. Parameter Estimates of the Multiple Regression of the Socio-economic Determinants of Building Technology Students' Achievement.

Variables	Linear	Semi-Log	Double-Log {a}
INTERCEPT	8.605 (4.211)**	10.943 (0.612)***	8.877 (1.517)***
Gender	1.862 (0.458)***	0.298 (0.303)	0.699 (0.303)**
Age	-21.308 (20.781)	-0.040 (0.010)***	-1.244 (0.282)***
Study Hours	1.264 (4.697)***	9.006 (6.570)***	1.062 (6.997)***
Parents' Education Background	1.747 (0.483)	0.102 (0.083)	0.001 (0.377)
Family Size	-3.795 (4.465)	-0.011 (0.028)	-0.452 (0.193)**
Family Type	-3.321 (-0.893)	0.209 (0.376)	0.005 (1.374)
Stipend per month (₦)	1.434 (0.457)***	0.068 (0.014)***	0.338 (0.032)***
Group Reading	0.283 (0.186)	0.118 (0.038)***	0.027 (0.009)***
Frequency of Parents' Visits	0.117 (0.120)	0.100 (0.060)*	0.068 (0.058)
Students Health Status	-2.025 (1.244)	-0.125 (0.055)**	-0.218 (0.060)***
No Credit Pass in SSCE/NECO	3.241 (1.393)**	3.387 (0.928)***	0.203 (0.065)***
Truancy	-1.381 (0.989)	0.320 (0.285)	-0.046 (0.024)*
Accommodation Status	-0.609 (0.251)**	-0.455 (0.205)**	-0.263 (0.189)
Involvement in Social Activities	-1.306 (1.021)	-0.215 (0.159)	-0.090 (0.063)
No of Observation	113	113	113
R^2	0.582	0.685	0.709
Durbin-Watson	1.865	2.001	2.130
F-value	9.152	21.374	27.888

Notes
*** denotes sig. at 1%;
** denotes sig. at 5%;
* denotes sig. at 10%.
Figures in parentheses () are standard errors; **{a}** = Lead equation;

Students' health status was highly significant but negatively related with academic achievement in Building technology at 1%. The negative implication of the relationship between health and academic achievement indicates that students with health challenges are more likely to have low CGPA than those that have no health challenges. This conforms with *a priori* expectation as health challenges could distract students' attention from learning and understanding which could culminate to poor achievement performance. The findings of this study on influence of health challenge of performance agreed with the result of Chude (2017) who found a significant but negative relationship between students with health challenges and academic performance. The coefficient of number of credit pass at SSCE/NECO was highly significant at 1% and positively related with academic achievement in Building technology students. The result shows that students with higher number of credit pass have higher CGPA at the end of their secondary year in Building technology. Of cause, this is expected transition from secondary school to higher institution. The findings corroborated that of Akenbor & Ibanichuka (2014) who found that students' achievement is significantly influenced by class size, entry requirement, access to functional library, semester duration, contact hours, and curriculum contents. Truancy was significant but negatively related with CGPA of Building technology students at 10%. The more the students stay away from class, the lower the academic achievement at the end of the session. The report of West Africa Examination Council (2014) ascertained that poor background of students and truancy constitute part of the factors that significantly influence achievement.

3.2 *Challenges Facing Building Technology Students in the Study Area*

The data in Table 3 presents the mean ratings of the students on the challenges they are facing in studying Building technology in colleges of education in North-central Nigeria. From the result, epileptic power supply to classroom and students' hostels for private study/reading (3.78), inadequate ICT facilities for effective learning of Building technology in the school (3.73), lack of functional Building technology workshop in the school (3.60), unequipped libraries in the school for relevant book consultation by students (3.56), lack or inadequate instructional materials (3.54) and poor classroom conditions in the school (3.52) had mean values that are within the boundary limit of 3.50 – 4.00 on 4-point rating scale. This indicates that the identified items are very serious challenges facing Building technology students. Poor attitudes of most students to learning of Building technology (3.44), too many number of courses in Building Technology offered by students (3.40), inadequate spacing between courses/paper during examination (3.33), inadequate exposure of students to practical works in Building technology (3.16), the use of ineffective teaching methods by most lecturers for instructional delivery (2.88) and short lecture periods on time table for Building technology instruction (2.75) had mean values that are within the boundary limit of 2.50 – 3.49 on 4-point rating scale. This indicates that the identified items are serious challenges facing Building technology students in North-central Nigeria.

The findings of this study agreed with Aremu & Sokan (2003) who found out that the students' factors of poor academic performance were poor study habits, psychological adjustment problems, lack of interest in school programme, low retention, association with wrong peers, low achievement motivation and emotional problems. Ogundele et al. (2014) expressed worries about the not-too-encouraging attitude of students which has been an impediment to their good academic performance in the country. In addition, Ong et al. (2010) who found that students' lack of financial support, absenteeism, truancy, use of local language in the classroom, lack of interest and joy in teachers' lessons and learning disability cause poor academic performance of students. Ogundele et al. (2014) also found that polygamous and large families contributed to poor academic performance of the students. This is because, parents' inability to provide breakfast, textbooks and basic school needs for their children, less interaction with children's teachers and less involvement in the Parents-Teachers Association (PTA) resulted in poor academic performance of students.

Poor attitudes of some lecturers of building technology to work (2.45), very short semester duration (2.39) and inadequate linkages between colleges of education and building industry (2.34) had mean values that are within the boundary limit of 1.50 – 2.49 on 4-point rating scale. This implies that the identified items are not serious challenges facing Building technology students in North-central Nigeria.

Table 3. Mean Ratings of the Challenges Facing Building Technology Students (n = 113)

SN	Challenges	Means	SD	Ranking Order
1	Epileptic power supply to classroom and students' hostels for private study/reading	3.78***	0.44	1st
2	Inadequate ICT facilities for effective learning of Building technology in the school	3.73***	0.48	2nd
3	Lack of functional Building technology workshop in the school.	3.60***	0.47	3rd
4	Unequipped libraries in the school for relevant book consultation by students	3.56***	0.52	4th
5	Lack or inadequate instructional materials	3.54***	0.45	5th
6	Poor classroom conditions in the school	3.52***	0.49	6th
7	Poor attitudes of most students to learning of Building technology	3.44**	0.48	7th
8	Too many number of courses offered by students	3.40**	0.55	8th
9	Inadequate spacing between courses/paper during examination	3.33**	0.48	9th
10	Inadequate exposure of students to practical works in Building technology	3.16**	0.51	10th
11	The use of ineffective teaching methods by most lecturers for instructional delivery	2.88**	0.50	11th
12	Short lecture periods on time table for Building technology instruction	2.75**	0.64	12th
13	Poor attitudes of some lecturers of Building technology to work	2.45*	0.47	13th
14	Very short semester duration	2.39*	0.52	14th
15	Inadequate linkages between colleges of education and Building industry	2.34*	0.50	15th

Notes
*** *Very Serious Challenges;*
** *Serious Challenges;*
* *Not Serious Challenges;*

4 CONCLUSION

This study therefore investigated socio-economic determinants of students' academic achievement in building technology using colleges of education in North-central Nigeria as case study. Based on the data collected and analysed, the study found that Building technology education in Nigeria is male dominated. The significant variables that influenced students' achievement in Building Technology were: gender, age, study hours, family size, stipends per month (₦), group reading, students' health status, number of credit pass in SSCE/NECO and truancy. The challenges facing Building technology students in the study area include epileptic power supply to classroom and students' hostels for private study/reading, inadequate ICT facilities for effective learning of Building technology in the school, lack of functional Building technology workshop in the school, unequipped libraries in the school for relevant book consultation by students, lack or inadequate instructional materialsand poor classroom conditions in the school among others.

The study recommended that; (1) efforts by government and other stakeholders in vocational and technical education in Nigeria should be geared towards bridging gender gap for increased enrolment of females in vocational and technical education, (2) students should be encouraged to engage in private study and group reading with peer in school to share academic ideas among themselves, (3) parents on their parts should ensure adequate provision instructional and other material resources including finance for their wards for effective learning and achievement in school, (4) the school management should make adequate provision of health facilities for care of students with health challenges that may impair their academic performance, (5) since number of credit pass at senior secondary level examination influence achievement in higher institutions, the school management should ensure strict compliance to standard on entry level required for admission into school, (6) the school management should made adequate provision of power supply to classroom and students' hostels for private for effective private study of students for higher academic achievement and, (7) there should be adequate provision of ICTs and other instructional materials in school libraries for quality learning in school.

REFERENCES

Abdullahi, H.A., Mlozi, M.R.S. & Nzalayaimisi, G.K. 2015. Determinants of students' academic achievement in agricultural sciences: A case study of secondary schools in Katsina State, Nigeria. *African Educational Research Journal* 3(1): 80-88.

Ademola, E.O., Ogundipe, A.T. & Babatunde, W.T. 2014. Students' enrolment into tertiary institutions in Nigeria: the influence of the Founder's Reputation – a Case Study. *Computing, Information Systems, Development Informatics and Allied Research Journal* 5(3): 55-64.

Akenbor, C.O. & Ibanichuka, E.A.L. (2014). Institutional factors influencing the academic performance of students in principles of accounting. *International Journal of Higher Education Management (IJHEM)* 1(1): 15-26.

Akinoso, S.O. 2017. *Correlates of some factors affecting students achievement in secondary school mathematics in Osun state, Nigeria.* An Unpublished manuscript, Department of Teacher Education, University of Ibadan, Nigeria.

Alufohai, P.J. 2016. School-based factors affecting senior secondary school students' achievement in English Language in Edo state. *European Journal of Research and Reflection in Educational Sciences* 4(9): 36-43.

Aremu, A.O. & Sokan, B.O.A. 2003. *A multi-causal evaluation of academic performance of Nigerian learners: Issues and implications for national development.* Department of Guidance of Counselling: University of Ibadan.

Chude, E.C. 2017. *Students' factors influencing their academic achievement in technical colleges in South-east, Nigeria.* An Unpublished Master's Thesis, Department of Vocational Teacher Education, University of Nigeria, Nsukka.

Diraso, D.K., Manabete, S.S., Amalo, K., Mbudai, Y.D., Arabi, A.S. & Jaoji, A.A. 2013. Evaluation of students' performance in technical and engineering drawing towards an effective career choice in engineering and technical and vocational education. *International Journal of Educational Research and Development* 2(4): 89-97.

Dokubo, I.N. & Deebom, M.T. 2017. Gender disparity towards students enrollment in technical education in Rivers state: Causes, effects and strategies. *International Journal of Research – Granthaalayah* 5(10): 1-10.

Egboh, S.H.O. 2009. Strategies for improving the teaching of science, technical and vocational education in schools and colleges in Nigeria. *Paper Presented at the one day Intensive Nationwide Training/Workshop Organized by the Centre for Science, Technical and Vocational Education Research Development,* Jos and Proprietors of Private Schools in Delta State held at College of Education: Warri.

Egun, A.C. & Tibi, E.U. 2010. The gender gap in vocational education: Increasing girls access in the 21st century in the Midwestern states of Nigeria. *International Journal of Vocational and Technical Education* 2(2): 18-21.

Federal Republic of Nigeria. 2014. *National Policy on Education.* Lagos: Nigerian Educational Research and Development Council (NERDC).

Igberadja, S. 2016. Effects of teachers' gender and qualification on students' performance in vocational technical education. *Journal of Technical Education and Training (JTET)* 8(1): 34-42.

Jeynes, W.H. 2002. Examining the effect of parental absence on the academic achievement of adolescents: The challenge for controlling family income. *Journal of Family and Economics Issues* 23(2): 189-210.

Kehinde, T.M. & Adewuyi, L.A. 2015. Vocational and technical education: A viable tool for transformation of the Nigerian economy. *International Journal of Vocational and Technical Education Research* 1(2): 22–231.

Kolawole, C.O.O. & Dele, A. 2002. An examination of the national policy of language education in Nigeria and its implications for the teaching and learning of the English Language. *Ibadan Journal of Education Studies* 2(1): 12-20.

Manyindo, B. 2005. Pilot project on co-operation between educational institution and enterprise in technical vocation education in Ganda, *presented at the UNESCO Seminar*, Breda: Netherlands.

Maurice, J.E., Asu-nandi, P.B., & Ntui, E.A. 2017. Entrepreneurial skills development and the building technology curriculum. *Niger Delta Journal of Education* 3(1): 600–607.

May, A., Ajayi, I.A., Arogundadade, B.B & Ekundayo, H.T. 2007. Assessing realities and challenges of technical education in Imo state secondary school education system in Nigeria. *Journal of Educational Administration and Planning* 7(3).

National Board for Technical Education (NBTE). 2012. *Vocational and technical education in Nigeria.* Kaduna: National Board for Technical Education.
Ogundele, G.A., Olanipekun, S.S. & Aina, J.K. 2014. Causes of poor performance in West African School Certificate Examination (WASCE) in Nigeria. *Scholars Journal of Arts, Humanities and Social Sciences* 2(5B): 670-676.
Ogunsola, O.K., Osuolale, K.A. & Ojo, A.O. 2014. Parental and related factors affecting students' academic achievement in Oyo state, Nigeria. *International Journal of Social, Behavioral, Educational, Economic, Business and Industrial Engineering* 8(9): 3129-3136.
Okoro, C.E. 2010. *Innovations in building technology and curriculum revision needs for building construction programmes of Colleges of Education (Technical).* An Unpublished Masters Project, Department of Vocational Teacher Education (Industrial Technical Education) University of Nigeria, Nsukka.
Okoye, R. & Arimonu, M.O. 2016. Technical and vocational education in Nigeria: Issues, challenges and a way forward. *Journal of Education and Practice* 7(3): 113–1118.
Ong, L.C., Chandron, V., Lim, Y.Y., Chem, A.H. & Poh, B.K. 2010. Factors associated with poor academic achievement among urban primary school children in Malaysia. *Singapore Medical Journal* 51(3): 247-252.
UNESCO. 2001. *Technical and vocational educational training.* Paris: United Nations Educational Scientific and Cultural Organization (UNESCO).
Unongo, J. 2015. *Comparative effects of dialogic teaching and coaching instructional strategies on students' performance, interest and retention in motor vehicle mechanic works in technical colleges in Benue state.* Unpublished Ph.D Thesis,, University of Nigeria, Submitted to the Department of Vocational Teacher EducationNsukka.
West Africa Examination Council [WAEC]. 2014. *Statistics of entry performance in WAEC examinations highlights of results released.* Abuja: Nigeria.

TVET Towards Industrial Revolution 4.0– Hazirah Noh@Seth et al. (eds)
© 2020 Taylor & Francis Group, London, ISBN 978-0-367-24273-2

Teaching of entrepreneurship skills as the means to sustainability

Seriki Mustapha Kayode
Kwara State College of Education (Technical), Lafiagi, Nigeria

Mohd Khata Jabor, Nornazira Suhairom, Nur Husna Abd Wahid & Zakiah Mohamad Ashari
School of Education, Faculty of Social Sciences and Humanities, Universiti Teknologi Malaysia, Johor, Malaysia

ABSTRACT: Developing students' entrepreneurial skill cannot be overemphasized, as the so-called ready-made jobs are not even available for graduates. This research study looks into how entrepreneurship can be integrated into not only higher education systems of learning but also starting from the post-primary level of education, and what teaching approach will be suitable for this integration. However, literatures have been written by experts on the entrepreneurial skills, based on how the curriculum could accommodate it, the change in the pedagogical approach and on the offering of the subject/course by non-science-oriented students at all levels of our education system. It was on this note that the researcher makes meta-analysis of the literature through library sources on what benefits entrepreneurial skills can offer as the solution to the rate of unemployment, the role of the teacher and the need to promote entrepreneurial skills. In conclusion, the importance of entrepreneurship in the development of a nation cannot be underestimated. It is recommended that student-centered teaching and learning situation should be adopted for teaching entrepreneurship.

1 INTRODUCTION

Much research has been carried out on the benefit of being an entrepreneur, hence the call for its integration for students, right from their post-primary education. Meanwhile, it is worth knowing the pedagogical approaches being employed by teachers in achieving the objectives of entrepreneurship courses/subjects in our schools. Presently, throughout the world, skills acquisition is being clamored for, and its importance cannot be overemphasized. These skills are quite different and in parallel with science and technological skills development. The skills are sometimes referred to as soft skills, critical thinking, or decision-making, and are generally composed of individual characteristics or the ability to change with the situation and time. The Partnership for 21st Century Skills (2008) opined that these skills are endowed with problem-solving ability, creative thinking, communication and collaboration skills, and entrepreneurial thinking, as well as the ability to solve problems in an extraordinary way. The entrepreneurship skills include being creative and innovative, being financially disciplined, taking opportunities and getting the best out of a situation with limited risk.

These are the sets of skills that need to be introduced into the school curriculum right from the secondary level of education and its equivalent and up to the post-secondary school levels of the education system.

2 LITERATURE REVIEW

Much literature has been written by educational experts on entrepreneurship, ranging from advocacy for entrepreneurship inclusion in the curriculum for all students, write-ups on pedagogical approaches by teachers, and offerings of entrepreneurship for non-business-oriented students in our education system at all levels. Onyeniyi (2003) stated that the word entrepreneur derives from the French word 'entrepreneur', which means undertake, which is the process of supplying goods or services to market for profitable purposes. Entrepreneurship involves risk-taking and organizational management (Leebaert, 1990). Butter (1990) described an entrepreneur as one who takes risks in business enterprise. The Finnish Enterprise Agencies (2014) explained the concept of entrepreneur in relation to education; this includes entrepreneur, entrepreneurship, entrepreneurship education as well as enterprise. A profit-minded individual who performs business activities, either in a group or alone, and also the person or group of people who help others who engage in taking risks to certain extents, is referred to as an entrepreneur, while according to Bozkurt (2011), enterprise consists of an economic unit under an entrepreneur as a body of profit-makers. A new opportunity created by an entrepreneur in order to make profits is entrepreneurship (Fisher & Reuber, 2010).

In relation to education, Heinonen and Poikkijoki (2005) described entrepreneurship education as the activities of inculcating in the individual the knowledge and understanding of enterprise and entrepreneurship. The knowledge includes, business, risk-taking, innovation, being creatively minded, capability of acting independently, adapting to new conditions, taking opportunities, teamwork, self-confidence and time management (California Department of Education, 2013; Güven, 2009; McKinney, 2013; National Content Standards for Entrepreneurship Education, 2004).

Researchers and teachers, especially science-oriented teachers, perceived the importance entrepreneurial education can offer science students by integrating entrepreneurship into science education. Koehler (2013) emphasizes that the possibility of innovative creativity can be achieved if entrepreneurial education is integrated into science education (Bacanak, 2013; Bolaji, 2012, Deveci & Cepni, 2014; Ugwu et al., 2013). Entrepreneurial education in science education should be practically oriented, and science teachers must have a positive perception that entrepreneurship education will be successful if integrated and seen as a student-centered subject/course. Studies have shown positive results as to how a science education demonstrates a better understanding of the application of entrepreneurship education (Adeyemo, 2009; Abdu, 2011; Achor & Wilfred-Bonse, 2013; Deveci, 2016; Buang et al., 2009; Ezeudu et al., 2013; Ejilibe, 2012; Nwakaego & Kabiru, 2015; Ugwu et al., 2013).

2.1 *Entrepreneurship education*

Education is the process by which individuals undergo acquisition of knowledge, skills, attitude and the ability to live effectively in society (Okeke, 2007). Education prepares individuals in the society for better living, to be enterprising, employable, and employers of labor, as well as an entrepreneur. To this end, the integration of entrepreneurship education in science and technology, and other fields of study not limited to social science classes, will go a long way to providing a sustainable living to graduates. Aguele and Agwagah (2007) emphasized that this will bring about economic and social development and improve the welfare of the recipient. Bolarinwa (2001) defines entrepreneurship education as education that provides training, experience and skills that are suitable for entrepreneurial endeavor. Entrepreneurship education can be used as a means of preparing graduates to be self-reliant, competent and knowledgeable, and to acquire skills. Entrepreneurship is the subject/course that offers students opportunities to anticipate and respond to changes (Ashomore, 1989). Iloputaife (2002) stated that the functionality in education (entrepreneurship in science and technology education) would serve to:

1. Identify students that possess entrepreneurial traits;
2. Motivate and develop students for launching and managing their own small-scale business enterprise;
3. Create necessary awareness and motivation in students for promoting self-employment and alternatives to wage empowerment.

2.2 *Advantages of entrepreneurship education*

Bolarinwa (2001) stated the following as part of the advantages of entrepreneurship education:

1. Entrepreneurship education helps students to acquire basic and necessary operational knowledge about the business environment.
2. It develops occupational knowledge, job skills and work experience in the students.
3. It provides the opportunities for students to invest in the early stage of their life into more profitable business than their peers.
4. It helps to reduce the dependence on white-collar jobs and to become self-employed by owning a business themselves.

2.3 *The role of teachers in teaching entrepreneurship*

Teachers are known as a role model and an implementer of curriculum. Adeyemo (2009) believes that for a teacher to be successful as an entrepreneurship instructor, the following entrepreneur skills/traits are essential:

- The requisite technical knowledge and expertise.
- The wisdom to seek out and listen to the advice of those who know what they don't.
- Ability to learn from their mistakes.
- Self-discipline.
- Ability to make quality decisions: the talents to analyze complex situations and draw conclusions that will enable the business to succeed.
- Hardworking: being capable of doing the work and enjoying it.
- Concentration: sticking it out through distractions to get the work done.
- Technical ability: the expertise to produce the goods and services of your business.
- Communication skills: the ability to express yourself and to understand others so that ideas can be shared.
- Motivation: the mental and physical drive to succeed, to accomplish chosen tasks on your own terms.
- Organizational skills.
- Decision-making skills.
- Financial skills.
- Student-management skills.
- Publicity/marketing skills.
- Supervision/management skills.

All of these enable teachers to have the broadest array of entrepreneurial skills, which are needed in the present market. Qualities expected in an entrepreneurship teacher include managerial skills, business skills, personal attributes and educational skills. Adeyemo (2009) also identifies the 'must-have' skills for the teacher of entrepreneurial education, including:

- Sales and marketing skills: an approach on how to get your target audience and an understanding of the needs of the people. These skills help the teacher to inculcate in the student the best marketing strategy.
- Financial know-how: management of finance is important and it will be appropriate if the teacher has financial discipline that can be transferred to their student, so that the students learn about how to manage a business without being extravagant.
- Time-management skills: as a teacher of entrepreneurship education, time management should be a priority to you, as time wasted cannot be regained. Time is money and time is precious.
- Administrative skills: an entrepreneur must be sound in management and in an administrative role; likewise, this skill is also important for teachers in entrepreneurship education.

2.4 *Pedagogical approaches needed in promoting entrepreneurship skills*

Discussion/case study method: Adeyemo (2009) applauds this method as it gives the teacher opportunities to present a case or scenario to the students to debate. Using a structured question, the teacher guides the discussion to avoid distraction and diversion from the case study at hand, so that the students will reach a constructive conclusion to the case. This strategy is said to be suitable for students in developing entrepreneurial skills.

The benefits of a case study as a pedagogical approach is that it gives room to evaluate all reasonable inferences; it provides an open floor for students to come up with ideas, it allows alternative interpretation of points made, it explains evidence better, it is simple and easy to construct, it covers a wider spectrum of the study at hand (entrepreneurial skill acquisition), it accepts new ideas, it allows re-evaluation and assessment of situations and re-orients interest in entrepreneurship skills.

Question method: The questioning method, according to Adeyemo (2009), is classified into two parts: reciprocal peer questioning and reader's questions. The former allows the teacher to display a list of questions, while students must write questions about the lecture materials, and then ask each other questions in a small group. The latter requires students to carry out assigned reading and during the class the students ask questions of each other as a discussion. Inline of this Grayson (1997) stated that posing strategic questions and helping students build on each others' ideas. The questioning strategy can also be referred to as conference-style learning, because the teacher serves as facilitator and the students read the materials thoroughly ahead of the class. The reading must be understood by students; the teacher does not remain passive, but will be active in helping and directing the section.

Ambiguity: This involves giving unclear material to students to brainstorm and, in the process, come to a structured conclusion. Scott (1998) advocates for more ambiguity in the classroom. In choosing the teaching methods, the following factors should be considered:

1. Teacher competency and experience;
2. The maturity and experience of the students;
3. Teacher aid availability (the instructional materials);
4. Time factors.

Meanwhile, the purpose of each method matters most in order to achieve the objective of the course or subject. Other non-business course/subject teachers can also adopt the following teaching and learning methods in developing entrepreneurial skills. Martin and Iucu (2014) listed the recommended teaching methods:

- Making use of active methods: that is, problem solving, simulation, role playing, discovery learning, etc.; this will contribute effectively to the development of responsiveness and help in building problem solving abilities in the student.
- Observation method: carrying out observation, developing business plans and portfolios, use of the Internet and educational software and teamwork in order to practice entrepreneurial skills. The observation method also includes monitoring and evaluation of the business.

3 OBJECTIVE OF THE STUDY

The objective of this study is to find out about the pedagogical approaches employed by entrepreneurial education teachers as a tool to make students self-reliant.

4 METHODOLOGY

This study adopts a systematic way to search through search engines, the Universiti Teknologi Malaysia (UTM) library database, where subscriptions were made to high-impact journals, such as *Science Direct* (Springer), *Web of Science* (Elsevier) and Google Scholar. Over 300,000 results were obtained when the keyword 'entrepreneurship education' was searched for, and keyword search also included 'teaching entrepreneurship education'. The results were too numerous for this study, with many articles not directly relevant to the purpose of the meta-analysis. Meanwhile, to find the most recently published articles, the search was limited to articles from 2009 to 2018. Later, the article search was structured to teaching of entrepreneurial skills, and to this end large numbers of irrelevant studies were waived. The search was concluded when a total of ten studies were selected for the meta-analysis. The abstract and full contents of these articles were read and the results were analyzed. This study focuses on students as well as the teachers in entrepreneurship. Table 1 describes the formulation of the study.

Table 1. Summary of reviewed articles.

Authors	Study objective	Methodology	Participants	Findings
Deveci, 2016	Investigate the concept of entrepreneurial characteristics and the perception of pre-service science teachers in the process of transferring the entrepreneurial character of the student.	Qualitative study	Students	The perceptions of pre-service teachers in relation to entrepreneurial characteristics where the student teachers are conscious of the need for active and creative ideas. Meanwhile, they are curious about the concept of entrepreneurship. It shows that the students have insufficient perceptions regarding the concept of entrepreneurship. The students are lacking the necessary knowledge of entrepreneurship.
Ezudu et al., 2013	To study the importance of Science, Technology and Mathematics (STM) entrepreneurship.	Descriptive study		The education policy on STM needs to be amended to accommodate entrepreneurship. Also, regular monitoring, supervision and evaluation of the program. The development of intellectual manipulative social and other skills that will ensure self-fulfillment and self-reliance of citizens. Equipping the STM laboratory, workshop, and provision of adequate funding.
Adeyemo, 2009	To illustrate the concept of acquiring entrepreneurial skills. A reorientation for classroom teachers in science education.	Descriptive study	Teachers	Acquisition of essential entrepreneurial skills are needed by science teacher as a tools and mean for improvement and this act as stimulating factor for the development of management competencies in teachers. Teachers as decision makers should understand while students have had problems in being entrepreneurial and what economic, educational and political changes in order to foster the development of entrepreneurial skills of students. Therefore, students in schools have to realize that in order to succeed in the work places of the future; they have to prepare themselves for the entrepreneurial path ahead regardless of their chosen discipline.

(Continued)

Table 1. (*Continued*)

Authors	Study objective	Methodology	Participants	Findings
Bacanak, 2013	The study was to determine the views of science and technology teachers about the effects of grade science and technology courses on students' entrepreneurship skills.	Qualitative (Interview)	Teachers	New behavioral change has to be created by a teacher; for this reason a student-centered method would assist the development of individual entrepreneurial skills.
Achor & Wilfred-Bonse, 2012	Examine the responsiveness of science teachers to global challenges and instructional demand of the 21st-century teacher education level in relation to entrepreneurial skills.	Descriptive study	Teachers	There is a need for proper re-direction of teachers' education to be more productive in terms of knowledge, skills, value and attitude to tackle the challenges of the new global rise in unemployed graduates.
Al-Atabi & Deboer, 2014	To study the effectiveness of traditional methods of teaching entrepreneurship.	Quantitative	Students	Massive Open Online Courses (MOOCs) offer a thinking framework, required by the brain to resist negative stimuli in students.
Mosey, 2016	To explore the entrepreneurial opportunities in technology, through teaching technology entrepreneurship.	Descriptive study		The new approach of teaching entrepreneurship on governing the teaching of technology entrepreneurship in higher institutions of learning.
Bindah & Magd, 2016	To examine and recommend structural conditions necessary and successful approaches in teaching entrepreneurial skills.	Descriptive study		Entrepreneurship should be introduced to all disciplines in higher institutions of learning, in order to create graduates that are self-employment oriented.
Martin & Iucu, 2014	Examine the need to develop entrepreneurship competencies of students from an educational science field.	Survey	Students	Change in the approach of teaching entrepreneurial skills is needed in the science field. The attitudinal change, soft skill levels, and behavioral change as a result of exposure the student faced during the experiential study was recorded in an entrepreneurship class.
Murah & Abdullah, 2012	To study the transformation that has been done on the teaching and learning process in technology entrepreneurship.	Observation	Students	Students exposed to the real world of entrepreneurship skills have changed mindsets. The teaching approach was changed from the traditional methods to student-centered and a rewarding result was achieved.

5 DATA ANALYSIS

It can be observed from the summary in Table 1 that different instruments were used by the authors of the selected articles. Thus, according to findings, the use of qualitative and quantitative approaches give an in-depth and exploratory result. The combination of the two instruments, which involve analysis and collection, according to Hanson et al. (2005) and Creswell (2012), can be said to be the gathering of facts.

The authors have different views and approaches to the teaching of entrepreneurship, with some of the authors calling for its integration into science, technology and mathematics. Others, like Adeyemo (2009), emphasize the benefit of entrepreneurial skills to students and advocate for science teachers to diversify and embrace entrepreneurship as a means to self-sustainability for students. Martin and Iucu (2014) and Mosey (2016) call for a new student-centered approach to the teaching of entrepreneurship, whereas Murah and Abdullah (2012) advocate mindset change and the exposure of students to real-world experiences. Creative thinking and employability skills should be the goal of the teacher as this was part of the benefit of the entrepreneurial objective; Table 1 clarified this, as authors like Deveci (2016) and Achor and Wilfred-Bonse (2013) placed importance on the production of sound graduates through entrepreneurship skill acquisition.

6 CONCLUSION

The importance of entrepreneurship in the development of a nation cannot be underestimated. This can be achieved through its integration in all fields of study at all levels of education. The implementation of student-centered teaching and learning will bring out self-sustainability in students during their stay at school and after their graduation in an entrepreneurial environment.

REFERENCES

Abdu, B. (2011). Chemistry education and entrepreneurial development in Nigeria: Issues and challenges. *Coconut, 4*(1), 107–114.

Adeyemo, S.A. (2009). Understanding and acquisition of entrepreneurial skills: A pedagogical re-orientation for classroom teacher in science education. *Journal of Turkish Science Education, 6*(3), 57–65.

Achor, E.E., & Wilfred-Bonse, K.U. (2013). The need to integrate entrepreneurship education into science education teachers' curriculum in Nigeria. *Journal of Science and Vocational Education, 7*, 111–123.

Aguele, L.I. & Agwagah, U.N.V. (2007). Female participation in science, technology, and mathematics (STM) education in Nigeria and national development. *Journal of Social Science, 15*(12), 121–126.

Al-Atabi, M., & Deboer, J. (2014). Teaching entrepreneurship using Massive Open Online Course (MOOC). *Technovation, 34*(4), 261–264. doi:10.1016/j.technovation.2014.01.006

Ashomore, M.C. (1989). Challenging creativity through entrepreneurship. In B.E. Lankard, *Keys to the future of small business. The vocational education/entrepreneurship match*. ERIC Digest.

Bacanak, A. (2013). Teachers' views about science and technology lesson effects on the development of students' entrepreneurship skills. *Kuram ve Uygulamada Eğitim Bilimleri, 13*(1), 622–629.

Bindah, E.V. & Magd, H.A.E. (2016). Teaching entrepreneurship in Oman: Successful approaches. *Procedia - Social and Behavioral Sciences, 219*, 140–144. doi:10.1016/j.sbspro.2016.04.055

Bolaji, O.A. (2012). Intergrating enterpreneurship education into science education: Science teachers' perspectives. *Journal of Science, Technology, Mathematics and Education (JOSTMED), 8*(3), 181–187.

Bolarinwa, K.O. (2001). Incorporating entrepreneurship education into business education curriculum: An equilibrium way for sustainable poverty alleviation in Nigeria. In *Proceedings of The 14th Annual Conference of the Nigerian Association of Teachers of Technology (NATT)* (pp. 152–156).

Bozkurt, Ç.Ö. (2011). *Dünyada ve Türkiye'de girişimcilik eğitimi: Başarılı girişimciler ve öğretim üyelerinden öneriler*. Ankara, Turkey: Detay Yayıncılık.

Buang, N.A., Halim, L. & Meerah, T.S.M. (2009). Understanding the thinking of scientists entrepreneurs: Implications for science education in Malaysia. *Journal of Turkish Science Education, 6*(2), 3–11.

Butter, R. (1990). *Artisans and entrepreneurs in the rural Philippines: Case monograph 2*. Amsterdam, The Netherlands: Vu University Press.
California Department of Education. (2013). *Common core state standards, for English language arts & literacy in history/social studies, science, and technical subjects, for California public schools kindergarten through grade twelve*.Published by CDE Press 1430 N Street, Sacramento, CA 95814-95901
Creswell, J.W. (2012). *Qualitative inquiry and research design: Choosing among five approaches*. Thousand Oaks, CA: Sage Publications.
Deveci, I. (2016). Perceptions and competence of Turkish pre-service science teachers with regard to entrepreneurship. *Australian Journal of Teacher Education, 41*(5), 153–170.
Deveci, İ., & Çepni, S. (2014). Entrepreneurship in science teacher education. Journal of Turkish Science Education, 11(2),161-188. doi: 10.12973/tused.10114a
Ejilibe, O.C. (2012). Entrepreneurship in biology education as a means for employment. *Knowledge Review, 26*(3), 96–100.
Ezeudu, F.O., Ofoegbu, T.O. & Anyaegbunnam, N.J. (2013). Restructuring STM (Science, Technology, and Mathematics) education for entrepreneurship. *US-China Education Review A, 3*(1), 27–32. Retrieved from http://files.eric.ed.gov/fulltext/ED539960.pdf
Finnish Enterprise Agencies. (2014). *Guide: Becoming an entrepreneur in Finland*. Publisher by Suomen Uusyrityskeskukset ry
Fisher, E. & Reuber, R. (2010). *The state of entrepreneurship in Canada: February, 2010*. Ottawa, Canada: Publishing and Depository Services, Public Works and Government Services Canada. Retrieved from https://www.ic.gc.ca/eic/site/061.nsf/vwapj/SEC-EEC_eng.pdf/$file/SEC-EEC_eng.pdf
Grayson, D.J. (1997). A holistic approach to preparing disadvantage student to succeed in tertiary science studies. Part II. Outcomes of the Science Foundation Programme (SFP). *International Journal of Science Education, 19*, 107–123.
Güven, S. (2009). New primary education course programmes and entrepreneurship. *Procedia-Social and Behavioral Sciences, 1*(1), 265–270. doi:10.1016/j.sbspro.2009.01.048
Güven, S. (2010). An analysıs of life science course curricula from the perspective of the entrepreneurship characteristics. *E-Journal of New World Sciences Academy (NWSA), 5*(1), 50–57.
Heinonen, J. & Poikkijoki, S. (2005). An entrepreneurial-directed approach to entrepreneurship education: Mission impossible? *Journal of Management Development, 25*(1), 80–94. doi:10.1108/02621710610637981
Hanson, W.E., Creswell, J.W., Clark, V.L.P., Petska, K.S. and Creswell, J.D. (2005). Mixed methods research designs in counseling psychology. *Journal of Counseling Psychology, 52*(2), 224–235.
Iloputaife, E.C. (2002). Science and technology education and poverty alleviation. In *The science teacher today*. Enugu State, Nigeria: Federal College of Education Eha-Amufu.
Koehler, J.L. (2013). *Entrepreneurial teaching in creating third spaces for experiential learning: A case study of two science teachers in low-income setting* (Doctoral thesis, University of Illinois, Urbana-Champaign, IL).
Leebaert, D. (1990). *Elements of entrepreneurship, prosperity* (Papers series). United States Information Agency.
Martin, C. & Iucu, R.B. (2014). Teaching entrepreneurship to educational sciences students. *Procedia - Social and Behavioral Sciences, 116*, 4397–4400. doi:10.1016/j.sbspro.2014.01.954
McKinney, S.W. (2013). *4 reasons entrepreneurship is crucial to a middle school education*. Momentum for Growth. Retrieved from http://blog.safeguard.com/index.php/2013/09/17/4-reasons-entrepreneurshipis crucial-to-a-middle-school-education/
Mosey, S. (2016). Teaching and research opportunities in technology entrepreneurship. *Technovation, 57–58*, 43–44. doi:10.1016/j.technovation.2016.08.006
Murah, M.Z. & Abdullah, Z. (2012). An experience in transforming teaching and learning practices in technology entrepreneurship course. *Procedia - Social and Behavioral Sciences, 59*, 164–169. doi:10.1016/j.sbspro.2012.09.261
National Content Standards for Entrepreneurship Education. (2004). *Preparing youth and adults to succeed in an entrepreneurial economy, accelerating entrepreneurship everywhere*. Retrieved from http://www.entre-ed.org/Standards_Toolkit/Helpful%20Downloads/NCSEE%20Website.pdf
Nwakaego, O.N. & Kabiru, A.M. (2015). The need to incorporate entrepreneurship education into chemistry curriculum for colleges of education in Nigeria. *Journal of Educational Policy and Entrepreneurial Research, 2*(5), 84–90.
Odubunmi, E.O. (1983). *The effect of socioeconomic background and teaching strategies on learning outcomes in integrated science* (Doctoral thesis, University of Ibadan, Nigeria).
Okebukola, P.A.O. (1984). *Effects of cooperative, competitive and individualistic laboratory interaction patterns on students' performance in biology* (Doctoral thesis, University of Ibadan, Nigeria).

Okeke, E.A.C. (2007). *Making science education accessible to all: An inaugural lecture of the University of Nigeria, Nsukka (UNN)*. Nsukka, Nigeria: UNN Senate Ceremonial Committee.
Onyeniyi, O.A. (2003). Promotion of entrepreneurship through vocational education in Nigerian Colleges of Education. *Oro Science Educational Journal (OSEJ)*, *2*(1&2), 156–158.
Partnership for 21st Century Skills. (2008). *21st Century skills, education & competitiveness: A resource and policy guide*. Tucson, AZ: Partnership for 21st Century Skills.
Scott, P. (1998). Teacher talk and meaning making in science classrooms: A Vygotskian analysis and review. *Studies in Science Education*, *32*, 45–50.
Ugwu, A.I., La'ah, E. & Olotu, A. (2013). Entrepreneurship: Performance indicators for innovative/skill acquisition: Imperative to science and technology education (STE). In *Proceedings, World Conference on Science and Technology Education, 29 September – 3 October, Sarawak, Borneo, Malaysia.*

TVET Towards Industrial Revolution 4.0– Hazirah Noh@Seth et al. (eds)
© 2020 Taylor & Francis Group, London, ISBN 978-0-367-24273-2

Learning influence factors on construction technology programs at vocational colleges in Johor

Sarimah Ismail, Nur Syafika Kamis, Nornazira Suhairom & Dayana Fazeeha Ali
School of Education, Faculty of Social Sciences, Universiti Teknologi Malaysia, Skudai, Johor, Malaysia

Arif Kamisan Pusiran
Faculty of Business, Economics and Accountancy, Universiti Malaysia Sabah, Kota Kinabalu, Sabah, Malaysia

ABSTRACT: The Construction Technology program curriculum at vocational colleges in Malaysia consists of 32 skill-based core courses. Each course has theoretical and practical components. However, the cumulative results of each course showed students got higher marks for practical components than theoretical ones. A preliminary study applying a focus-group data-collecting technique was conducted to identify factors influencing students' learning that caused low achievement in the theoretical component. Factors identified were class timetable, workshop facility and the teaching strategy of the lecturer. These factors have been used as research variables for this quantitative study, which has been conducted at five out of six vocational colleges in Johor that are offering the program. 140 of the first and second year students of the program have been selected randomly as the sample for the study. A set of questionnaires was developed as a research instrument and was verified by two experts. The validity value of the questionnaire was $\alpha = 0.866$. Data was analyzed using descriptive statistics of frequency, mean and standard deviation, which was assisted by computer software. The findings of the study showed that all research variables influenced students' learning of the theoretical components of Construction Technology courses at a high level (class timetable: mean = 4.20 and standard deviation = 0.921; workshop facility: mean = 4.27 and standard deviation = 0.872; teaching strategy: mean = 4.42 and standard deviation = 0.811). The overall mean level of the influence of these three variables was also high, with a mean value = 4.30 and standard deviation = 0.868. This study suggests the class timetable, the workshop facility and the teaching strategy of the lecturer should be improved to help students of these programs obtain high marks in the theoretical component of each course.

1 INTRODUCTION

Education plays an important role in producing knowledgeable, skilled and balanced human capital for the development of Malaysia. The balance and educated society of Malaysia can be achieved through a quality education system. To transform Malaysia to a developed country that is based on technology and skill services, it needs a balance of both skilled and semi-skilled human capital. The upgrading of vocational and technical schools to vocational colleges is seen as a platform in providing continuous semi-skilled workers to the industries in Malaysia. Rapid growth of the industry promises national economy development.

However, to get semi-skilled workers with high technical skill competency, knowledge and equipped with generic skills, the vocational colleges should provide a conducive learning atmosphere, complete facilities, competent lecturers and promote active learning participation for the students. According to Azizi and Gangagoury (2008), the effectiveness of the teaching and learning process depends on how teachers understand the learning styles and learning problems of their students. The teacher should be able to adapt their teaching strategy to suit with the learning style of the students. This indicates that teachers should have knowledge, skill and passion, not only in their field of study, but also in pedagogy.

1.1 *Background of problem*

The Construction Technology is one of the programs offered at six out of eight vocational colleges in Johor State. The program consists of 32 skill-based courses that have both theoretical and practical components. However, the cumulative results of each course showed that almost all students achieved high marks in the practical component rather than the theoretical component of the courses. A preliminary study was conducted to identify the factors influencing student learning of the Construction Technology program that lead to different achievements between the theoretical and practical components of the courses.

Three series of focus groups have been conducted and involved 15 students of the Construction Technology program to get rich data for the development of research variables. According to Sarimah and Mohd Shafie (2014), the interview data-collecting technique can provide rich data compared with data collected using a survey method. The findings of the preliminary study showed that the three major factors that influenced students' learning of the program were the class timetable, the workshop facility and the teaching strategy of the lecturer. Among the feedback received from the students who participated in the focus group were the following:

> *I cannot focus on the teacher's explanation during theory class in the afternoon session due to feeling sleepy and tired.*
>
> *I feel lost in theory class in the afternoon. I do not know what the lecturer's said in front of the class because at that time I am usually sleepy and tired. Having theory class in the timetable after having lunch is not suitable at all.*
>
> *Students have to share tools and equipment to do their project in the workshop. This has caused delay in completing the project.*
>
> *The workshop has many tools and equipment but the functioning ones are limited and the students have to take turns to use the tools. I think college needs to ensure tools and equipment quantity should be equal with the number of students so that the students do not have to share and wait.*
>
> *Workshop activity is fun because it is a hands-on approach. Theory class is different. It is terrible. The student just sits and listens to what the lecturer explains and then we are asked to take notes. It is really boring.*
>
> *Teaching strategy that involves class activity will be interesting and make the class lively.*

Based on the findings of the preliminary study, this quantitative study was conducted to identify the influence levels of the class timetable, workshop facility and teaching strategy of the lecturer on students' learning of the theoretical component of Construction Technology courses.

2 METHODOLOGY

This study applied a quantitative research design. Data was collected from 140 Vocational Certificate students of the Construction Technology program. The sample size of the study was determined by the following equation:

$$S = \frac{X^2 NP(1-P)}{d^2(N-1) + X^2 P(1-P)}$$

$$S = \frac{(3.841)(220)(0.5)(1-0.5)}{0.05^2(220-1) + 3.841(0.5)(1-0.5)}$$

$$S = 211.255/1.5077$$

$$S = 140.117 \text{ or } 140$$

where S = sample size; X^2 = confidence level (at 3.841); N = population size.

Those students were selected randomly among 220 students of five vocational colleges that offer the Construction Technology program in Johor State. A set of questionnaires was developed as a research instrument and was verified by two experts in the field of Construction Technology and pedagogy. A pilot study was conducted at one vocational college in Negeri Sembilan and involved 30 students of the same program. The result of the pilot study found that the overall validity value of the instrument was $\alpha = 0.868$, while the validity value according to research variables of class timetable, workshop facility and teaching strategy of lecturing were 0.86, 0.85 and 0.87, respectively. These values showed that items of the research instrument are acceptable and the instrument can be used for actual data collection. According to Creswell (2013), a validity value for a survey research instrument that is close to 1.0 indicates that the research instrument has a high validity, is considered very good and ready to be used for data collection. Data of this study was analyzed using descriptive statistics of frequency, mean and standard deviation, and was assisted by the Statistical Package for Social Science (SPSS) version 22.0 computer software.

3 FINDINGS AND DISCUSSION

The results of this study indicated that the mean level of the three research variables influenced students' learning of the theoretical component of the Construction Technology courses at a high level (class timetable: mean = 4.20 and standard deviation = 0.921; workshop facility: mean = 4.27 and standard deviation = 0.872; teaching strategy of lecturers: mean = 4.42 and standard deviation = 0.811). The overall mean for the influence level of the variables was also high, with a mean value = 4.30 and standard deviation = 0.868.

Table1. Level of learning influence according to variables.

Variables	Mean	Standard deviation	Level of influence toward learning in Construction Technology courses
Class timetable	4.20	0.921	High
Workshop facility	4.27	0.872	High
Teaching strategy of lecturer	4.42	0.811	High
Overall mean	4.30	0.868	High

3.1 *The influence of class timetable on the learning of the Construction Technology program*

Ten items were prepared to identify the influence level of the class timetable on students' learning of the Construction Technology courses. Those items were related to the impact of theory-based class that held in the afternoon session (between 2.00 o'clock to 5.00 o'clock) and the practice-based that involved workshop activities in the morning session (from 8.00

o'clock to 1.00 o'clock) on students' learning. The findings of the study showed that all of the items influenced students' learning at the high level. The overall mean value of this research variable was 4.20 with standard deviation of 0.921.

According to Zainuri (2014), the most suitable times to learn topics related to low cognitive levels, such as knowledge and understanding that involve memorizing, are between 7 o'clock and 9 o'clock in the morning. At these hours, the level of individuals' thinking is high because their mind is fresh, strong and energetic. This indicates that a morning session is suitable for learning theoretical-based topics that involve understanding and memorizing.

Zainuri (2014) also said that the hours between 3 o'clock and 5 o'clock in the afternoon are suitable for teaching and learning that involve active activities. This means the afternoon session is suitable for hands-on-based learning, like workshop activities. However, the finding of our study is contrary to the statements of Zainuri.

The implementation of a theory-based class in the afternoon session (2 o'clock to 5 o'clock) and practice-based (workshop activities) in the morning session (8 o'clock to 1 o'clock) have given negative impact to students' learning. Tiredness, because of long hours of workshop activities in the morning session, caused the students to feel sleepy and distracted their focus on the content of the lesson of the theory-based class conducted in the afternoon session.

This study suggested that the arrangement of an eight-hour class timetable at the vocational college (from 8 o'clock in the morning to 5 o'clock in the afternoon) should be based on the nature of the course. A morning session (before the lunch hour) is suitable for theory-based courses, while an afternoon session (after the lunch hour) is suitable for courses that involve hands-on activity. For the situations where lecturers have no option other than teaching the theory component in the afternoon session, a teaching strategy that involves the active learning participation of students, such as active learning, cooperative learning and problem-based learning, should be adopted. This is to project students' attention toward the content of the lesson.

3.2 *The influence of the workshop facility on the learning of Construction Technology*

Ten items were prepared to identify the influence level of the workshop facility on the learning of Construction Technology courses. Those item were related to the insufficient number of functioning tools, equipment and machinery when compared to the number of students using the equipment, as well as their impact on students' learning. This research variable has influenced students' learning at the high level, with an overall mean value of 4.27 and standard deviation of 0.872.

The findings of the study found that almost all Construction Technology workshops at vocational colleges of study had an insufficient number of functioning tools and equipment compared with the number of students using the tools. As a result, the students have to share the tools and equipment with other students. This has led to delays in completing their project, forcing the course lecturer to use video as an alternative to show how the tools and equipment function to run the project.

Handling tools, equipment and machines is part of the competencies as stated in the National Occupational Skill Standard (NOSS) of the program that the students need to achieve before they can be awarded the Malaysia Skill Certificate on completing the program. Lack of skill and experience in handling these facilities due to the vocational colleges having incomplete workshop facilities will cause a negative impact on students' competencies in tool and equipment handling, as well as their job in industry. According to Mohd Najib et al., (2019), the number of tools and equipment should tally with the number of students, to ensure the students get sufficient knowledge and skills of the program in which they enrolled.

There are 85 vocational colleges in Malaysia (Norsuhaila, 2018) that offer more than 20 skill-based programs to produce semi-skilled workers in an attempt to fulfill the need of industry and Malaysia's mission as a developed country. Those skill-based programs need a complete workshop facility to ensure effective learning and teaching is conducted to provide knowledge and skills to students, so that they can be semi-skilled workers after completing their program of study.

The vocational colleges are responsible for ensuring that every workshop has a complete facility. The Per-head Capita Grant (PCG) amounting to RM220 allocated to each student per course per year is available to help the vocational colleges to buy tools and equipment, as well as course raw material.

The low amount of PCG for vocational subjects has been debated since it was introduced in 2004 for the daily academic school curriculum. The low amount of PCG has forced teachers who teach the subject to use their own money and use PCG for other vocational subjects in order to cope with the cost of learning raw materials (Zainuddin, 2010).

The subsequent research of Sarimah and Farawahida (2009) once again found that with a PCG amount of RM220, vocational subject teachers were facing difficulty in providing raw learning materials, which has a negative impact on their teaching process.

Hadawi and Crabbe (2018) suggested that the allocation of PCG for every course needs to be revised every time the curriculum of the course is revised to ensure that the amount of PCG is in accordance with the market price of the learning raw material. This is to ensure that the learning process runs smoothly.

In the current situation, where the value of the Malaysian Ringgit is in decline, the curriculum has been revised and the market price of almost all the learning raw materials has increased, as have the tools and equipment, this study suggested that the Malaysia Ministry of Education (MoHE) should revise and increase the amount of PCG of RM220 for each student/course/year to help vocational colleges to plan their budget for the skill-based courses that have many tasks to fulfill, where each task requires its own learning raw material. This is to ensure that the vocational colleges manage to produce competent semi-skilled workers for the development of Malaysia.

3.3 *The influence of lecturer's teaching strategy on the learning of Construction Technology*

Ten items were developed to identify the influence level of the teaching strategy of lecturers in relation to learning Construction Technology courses. Those items were related to the teaching approach and teaching strategies applied in teaching theory-based courses of the program and its impact on students' learning. The findings of the study showed that the majority of lecturers applied a teacher-centered teaching approach that uses talk-and-chalk in delivering course content. This approach has influenced students' learning at a high level. The overall mean value of this research variable was 4.42, with standard deviation of 0.811.

In the 21st century, the teaching approach has shifted from teacher-centered to student-centered. The student-centered teaching approach that involves student active participation in learning is able to equip students with generic skills. For instance, cooperative learning has proven to guide students working in teams, increase levels of self-confidence and improve communication skills. Innovation-based learning educates students with creative thinking and problem-solving, life-long learning and entrepreneurship skills, as well as ethics and professionalism. Project-based learning trains students in time management and leadership skills (Sarimah et al., 2018).

Because strategies of student-centered teaching have many advantages, this study suggested that for the theory-based component of Construction Technology courses at the vocational colleges of study, lecturers of the courses adopt those strategies in their teaching. Furthermore, where classes for theory-based components of the course are conducted in the afternoon session, where students feel sleepy and tired, a student-centered teaching approach that requires students' active learning participation through active learning activities will help to keep them awake, stay focused and acquire an in-depth understanding of the content of the lesson.

Systematic and well-prepared student-centered learning activities are not only able to strengthening students' technical knowledge of the course, but also manage to equip the students with generic skills. This is to prepare the students with employability skills in order to get employed and develop in the workforce. Dahiru and Sarimah (2016) found that the generic skills and core skills of the technical content of the field of study formed part of the employability skills. Thus, to produce balanced semi-skill workers for industry in Malaysia and cross the globe, teaching and learning activities should involve those student-centered learning strategies.

4 CONCLUSION

The findings of this study found that the class timetable, workshop facility and teaching strategy of the lecturer exerted high influence on students' learning of the theoretical component of Construction Technology courses. The high influence of research variables measured might have contributed to low achievement of the theoretical component in comparison to the practical component of the Construction Technology courses.

This study suggested that the class timetable committee at vocational colleges in Johor State should arrange the theoretical components of the Construction Technology courses and other courses that are also theory-based in the morning session. In the situation where the theory-based class has to be conducted in the afternoon session, a teaching strategy that involves students' active learning participation is strongly encouraged to be adopted to ensure the students give full engagement in the learning process. The advantages of the student-centered teaching approach are an in-depth understanding of course content, long-lasting memorization of content, and improvement of generic skills, which have all received high attention in the literature.

A sufficient number of functioning tools, equipment and machinery relative to the number of users, as well as a conducive workshop environment, are among the facilities that should be available at vocational colleges. Complete workshop facilities promise high experience and skill of students in handling the tools and machineries. Regular maintenance of the tools, equipment and machinery should be carried out to keep the facility in a good condition.

Practice-based component of the Construction Technology courses applied project-based learning that involved students in learning actively. As a result, all of the students got high marks for that component. The similar 21st teaching strategy that involves student active participation in learning should be adopted in teaching the theoretical component of the courses. This is to attract students' attention to learn and participate actively in learning the course content. The improvement of these three research variables is seen to help the students to increase their mark in the theoretical component of the Construction Technology courses and be knowledgeable semi-skill workers for Malaysia.

REFERENCES

Azizi, Y. & Gangagoury. (2008). *Factors influencing teaching and learning living skill at Sekolah Rendah Jenis Kebangsaan Tamil in Johor Bahru* (Final year project, Universiti Teknologi Malaysia, Johor, Malaysia).

Creswell, J.W. (2013). *Research design: Qualitative, quantitative and mixed methods approaches.* Thousand Oaks, CA: Sage Publications.

Dahiru, S.M. & Sarimah, I. (2016). Employability skills integration in university TVET programs: A strategy for reducing unemployment rate among graduates in Nigeria. *Journal of Applied Sciences & Environmental Sustainability*, *2*(4),44–55.

Hadawi, A. & Crabbe, M.J.C. (2018). Developing a mission for further education: changing culture using non-financial and intangible value. *Research in Post-Compulsory Education*, *23*(1),118–137.

Mohd Najib, A. K., Sarimah, I. & Zaharah, J. (2019). Workshop Facility and Per-head CapitaGrant of PVMA Automotive Program in Malaysia, *Educational Initiatives Research Colloquium Proceeding*, 202–206.

Mohd Najib, A. G. (1999). Educational Research Methodology. Skudai: Universiti Teknologi Malaysia Press.

Sarimah, I. & Farawahida, Y. (2009). Constraint in implementing vocational subject at secondary schools in Johor State. In *Proceedings of National Seminar of Technical and Vocational Education* 2009, *28–29 July 2009, Malaysia Teacher Institute of Technical Education Campus.*

Sarimah, I. & Mohd Shafie, M.S. (2014). Attracting and sustaining customers through antecedent satisfaction experience. In *Proceedings of the 2nd International Hospitality & Tourism Conference 2014 (IHTC2014) on Theory and Practice in Hospitality and Tourism Research, UiTM Hotel Penang*, 2-4 September (vol. 1, pp. 405–409).

Sarimah, I., Aede, H.M., Mohd Fa'iz, A. & Zairil, I.R. (2018). Innovation-based learning conceptual model. *Turkish Online Journal of Design, Art and Communication*, September 2018, Special Edition, 1697–1706.

Norsuhaila, T. (2018). *Workshop Management at Excellent Vocational Colleges in Johor* (Final year project, Universiti Teknologi Malaysia, Johor, Malaysia).

Zainuddin, A.B. (2010). *Teaching approach and the impact towards students achievement in living skills subject at Sekolah Menengah Kebangsaan Senai* (Final year project, Universiti Teknologi Malaysia, Johor, Malaysia).

Zainuri. (2014). Factors influencing time management of school student in Slim River District (Final year project, University Pendidikan Sultan Idris, Tanjung Malim, Malaysia).

TVET Towards Industrial Revolution 4.0– Hazirah Noh@Seth et al. (eds)
© 2020 Taylor & Francis Group, London, ISBN 978-0-367-24273-2

Employability skills of higher education graduates: A review and integrative approach

Rufus Sunday Olojuolawe
Department of Technical and Vocational Education, College of Education, Ikere-Ekiti, Nigeria

Nor Fadila Amin, Adibah Abdul Latif & Mahyuddin Arsat
Faculty of Social Sciences & Humanities, Universiti Teknologi Malaysia, Malaysia

ABSTRACT: Noting the importance of employment to graduates across disciplines, this study considers the best approach to studying employability skills in Higher Education research. Employability skills faces challenge in terms of appropriate methodology due to the different perceptions of the major stakeholders. This has brought about differences in the opinion of researchers on how best to approach the study of employability skills. Based on the review of previous studies in employability, a search string of journal articles was conducted using Scopus data set. The search spanned between 2013 and 2016 with a retrieval of 118 publications from the database. The number was reduced to 61 by automatic refinement. Further filtering reduced the publications to 42 for full study. Findings shows that there was no uniform methods of conducting research in employability. The study suggested a uniform way of conducting employability skills research based on the aggregate methods found in the study.

1 INTRODUCTION

1.1 *Overview of the study*

The acceptance of employability skills into the mainstream of tertiary institutions seems to be gaining considerable acceptance especially, among the developing countries. The growing interest is due to a large number of graduates from different disciplines who have remained jobless after leaving school. This has attracted interest from policymakers, employers, and educational institutions. Employability skill is a concern with the process of securing, advancing and sustaining a job for the economic benefits of the individual and the society at large.

Employability skills cover all educational courses. It is a lifelong process (Dacre Pool and Sewell, 2007). The increasing rate of unemployment is a source of worry to stakeholders (Asaju, 2014). Many of the graduates are unemployed because they lack the skills for current jobs (Uddin, 2013). It has been reasoned that the possession of hard skills alone is no longer fashionable for 21st-century jobs (Dean, 2017).

The increase in the rate of joblessness has led to the concern about the effectiveness of the curriculum being used by the Colleges (Asaju, 2014). Thus, emphases should be on the functionality of the school curriculum (Amusan *et al.*, 2016). The possession of academic qualification is no longer sufficient for today's labour market (Tymon, 2013). The nature of curricula offering provides evidence of skill mismatch between the school and the employers (Bowman *et al.*, 2016).

Noting that no single research approach is sufficient for a phenomenon like employability, the combination of approaches or designs would be of advantage for the improvement of graduate's employability skills.

Added to the serious issue of unemployment is the emergence of technological changes in 21-century (Mansour and Dean, 2016). This necessitated the need for looking into how best to arrest the drift into the unemployment market (Cardoso, 2014). Badu-Nyarko (2013) indicates that it is the duties of the higher institution to ensuring that relevant researches are conducted to permanently solve the problem.

However, there are challenges to researching employability skills due to diverse interpretations of the concepts. The definition of employability varied with researchers and changed in accordance with political discourse (Madar and Buntat, 2011). Lawal (2013) opined that the concept of employability skills may be subjective in terms of individual contextualization. Hillage and Pollard (1998) propose that employability is a construct of capitalism, and forms part of public discourse influencing individuals' cultural, social and vocational experiences.

This paper examines some of the methodological challenges faced by employability skills research. In order to improve the method used in the field. Suggestion for the integrative methodological framework from which future research projects may draw insights was made. Hence, a way to enhancing research in employability skills. Researchers should be more concerned with the learning outcome that is related to graduates usability to society.

This paper is a meta-analysis of the methods applied instant employability skills literature, emphasizing the strengths and weaknesses of the current research in the field. Then, an outline follows an integrated methodological framework for researching employability skill courses and programmes, before a discussion concludes the paper.

1.2 *Search strings*

A peer-reviewed of journal articles published between 2013 and 2016 was used for this study. The aim was to have an insight into the existing methods of conducting research in employability skills. The question that guided the study is concerned with assessing the method used to study different aspects of employability skills. To ensure consistency in the search and collection of sources for the review, Scopus citation index services online was used. The database allows researchers to compose explicit strings to identify potentially relevant publications across a large number of collections in different subject areas as shown in the framework.

Keywords on the subject were carefully composed into search strings as shown in Table 1.

Table 1. Search strings.

Main search: Employability
Sub search 1: Employability skill + Technical Education
Sub search 2: Employability skills + Vocational Education
Sub search 3 Employability skills + Technical and Vocational Education

2 METHODOLOGY

2.1 *Search strings*

The search framework on the subject – Employability in Vocational and Technical Education is shown in Figure 1.

Based on the search string a total number of 118 publications was retrieved from the database. The Scopus citation index services analytical tool was used to analyze the result of the search to reflect the publications by authors and regions of the world. This number was pruned down by setting scope for the search strings. Inclusions and exclusions were introduced as criteria for articles to be selected for review as follows:

2.2 *Inclusion and exclusion*

In order to narrow down our search for relevant review papers, publications with the following were chosen from the search engine: (i) papers published between 2013 and 2016, (ii) papers on social sciences, computer and engineering, (iii) documents that are either journal articles or conference papers, (iv) papers that discuss employability in technical or/and vocational education;

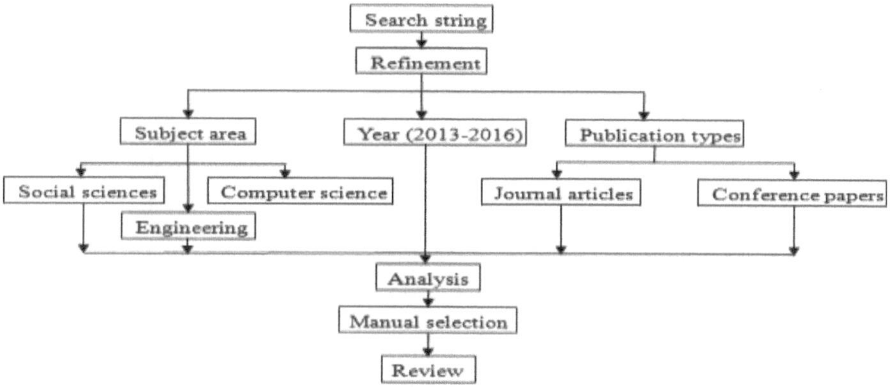

Figure 1. Employability in Vocational and Technical Education search framework.

and excluded the following publications (i) review papers, articles in press, book chapters, and (ii) publication not written in English. The number was reduced to 61 records by this automatic refinement.

Furthermore, the filtered literature was critically looked into through manual selection that was based on their relevance to the topics and the contents of the abstracts.42 publications were finally selected for the full-scale study. During the study, a meta-analysis table was formulated to capture data on the principal author, document type, year of publication, region, publication source (journal or conference paper), subject area and the technique used for analysis (qualitative, quantitative or mixed mode).

2.3 *Analysis and discussion of results*

If In analyzing the result of the systematic review, charts were generated for a pictorial representation of the results of the study. In generating the charts, the emphasis was laid on document type, a region of publication and subject area. The results of these are respectively shown in Figures 2 – 4. Also, the method adopted by each author in discussing employability in technical and vocational education was paid careful attention to with the aim of looking at their appropriateness for the study under review as shown in Table 2. Table 3 shows the list of journals from which the publications reviewed were drawn and their region.

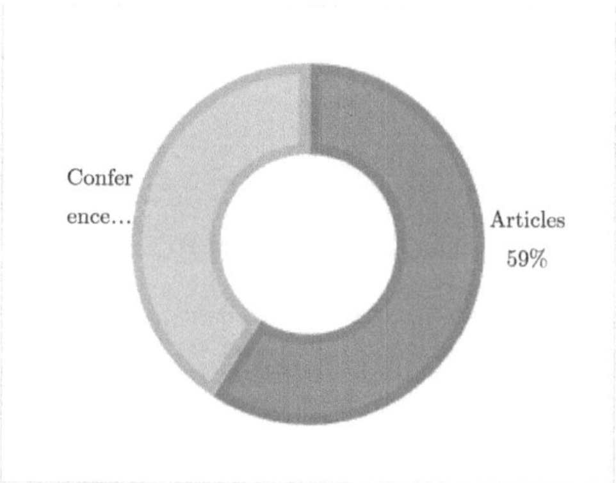

Figure 2. Distribution of publications by document type

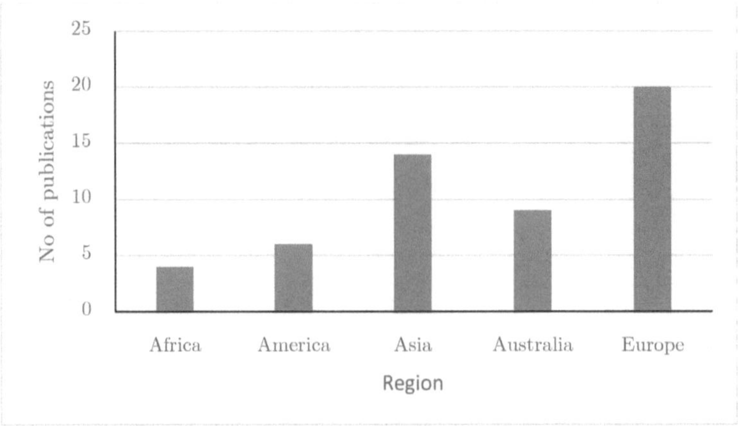

Figure 3. Distribution by region of publication.

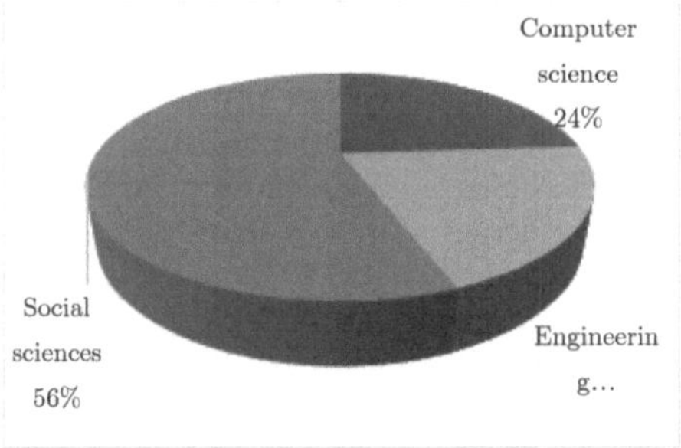

Figure 4. Distribution of publications by subject area.

Table 2. Studied methodologies.

Design/Technique	Total	Authors
Qualitative	4	(Gekara and Snell, 2013; Backa and Wihersaari, 2014; Kumar, Garimella and Nalla, 2014; Wedekind and Mutereko, 2016).
Quantitative	9	(Husain *et al.*, 2010; Boahin and Hofman, 2013; Družovec *et al.*, 2013; Rahim Bakar, Mohamed, and Hamzah, 2013; Saad *et al.*, 2013; Dania, Bakar and Mohamed, 2014; McDermott, Bass, and Lalchandani, 2015).
Mixed	2	(Renuga and Ezhilan, 2014; Ho *et al.*, 2016).
Nil	7	(Khlaifat and Qutob, 2013; Jayaram and Engmann, 2014; Mishra, 2014; Sgouropoulou and Georgouli, 2014; Soler and Andersson, 2014; Zeti and Faieza, 2014; Craft *et al.*, 2015; Mazhar and Arain, 2015; Bartlett and Pagliarello, 2016).
Others	10	(McDermott *et al.*, 2015; Chan, Lin and To, 2016; Davishahl and Swisher, 2016; Patro and Sahoo, 2016; Pieterse and van Eekelen, 2016; Santos *et al.*, 2016; Souto-Otero and Shields, 2016).

Table 3. Overview of Journals retrieved List of journal.

1. Journal of Vocational Education	2. Proceedings Frontiers in Education Conference
3. European Journal of Education	4. International Education Studies
5. On the Horizon	6. Procedia Engineering
7. Prospects	8. Anthropologist
9. Applied mechanics and Metrics	10. British Journal of Educational Studies
11. Communication in Computer and Information Science	12. Development of Southern Africa
13. Engineering Education	14. European Societies
15. Global Journal of Engineering Education	16. Higher Education Skills and Work-based Learning
17. IEEE International Professional Communication Conference	18. International Journal of Lesson and Learning Studies
19. International Journal of Electronics Commerce Studies	20. International Journal of Engineering Education
21. International Journal of Innovation in Science and Mathematics Education	22. International Journal of Technology Enhanced Learning
23. International Journal of Training Research	24. International Review of Education
25. Journal of Curriculum Studies	26. Journal of Engineering Design
27. Journal of Geography in Higher Education	28. Journal of Higher Education Policy and Management
29. Journal of Universal Teaching and Learning Practice	30. Research in Comparative Studies and International Education
31. Research in Post -Compulsory Education	32. Software Engineering Education Conference Proceedings

3 FINDINGS

3.1 *Findings of the literature review*

The findings of the review show the spread and emergence of publications on employability skills across regions and journals. The findings also revealed the different approaches used by authors to analyze their data.

The number of publications as shown in Figure 3 attests to the acceptance of employability skills across regions. Many of the papers are found in reputable international journals. This could also be ascribed to the strength and quality of the papers.

The review shows that employability skills research is embraced more in the other regions of the world than the African Continent. Employability skills enable an individual to secure employment and advance in it (Jackson, 2014). Consequently, the high rate of unemployment among youths in Africa may be significantly related to the weak acceptance of employability skills as against the regions of the world. Specifically, on the types of methods used in research on employability skills, the review shows that different types of methods were used by researchers. There is no coherence in the approaches adopted by different researchers. Table 2 shows the distribution of qualitative, quantitative, mixed methods and other approaches. A quantitative method was commonly used based on the findings of the study. This may be connected with the nature of the field of employability which gives room for direct observation of the participants to be able to give generalization (Creswell, 2013). The findings of the study reveal that the majority of the studies used a single methodological approach. The approach use was diverse and varied with different researchers. Considering the importance and objective of employability skills, it is important to look at the inherent advantages of using both qualitative and quantitative research methods in addressing issues in employability skills in addressing a single study. This would help in overcoming the demerits of both methods and combined the advantages. Table 4 gives the details. The integration of both methods will enhance the depth and eliminate bias.

Table 4. Merits and Demerits of Quantitative and Qualitative Approach.

Quantitative (Merit)	Qualitative (Merit)
– Description and explanation of process and change.	– Measuring variance and correlation
– Theory building and generalization sensitive to context.	– Possibility of statistical generalization
– Ability to integrate multiple perspectives	– Objective methods of analysis – Theory testing
Weakness	Weakness
– Inability to generalize to populations	– Limited ability to capture process and change
– Limited ability to test the theory	– Difficulty in establishing linear causality
– Over-reliant on the subjective judgment in analysis	– Lacking sensitivity to context.

3.2 *Conclusion and recommendation*

Employability skills are very important to all disciplines from engineering, medicine, architecture, management, and social sciences. This is because of the changing demands from the employers occasioned by the growing and changing ways of doing things in the industries including taste and demand. A number of researches by researchers have reached a very good pedestal especially in Europe where most of these journals come from. Findings of this study show that African researchers need to more in employability research. The number of journals coming from the continent of Africa is considerably compared to other continents.

Finally, we recommended the integration of mixed method for approaching employability skill studies in technical education because it lends itself for in-depth analysis of the phenomenon. This will practically help in identifying the root causes of unemployment in most countries and the best approach to combat the problems. This method would be most beneficial to the employers, academic institutions and the society at large. Thus, the weakness caused by a single research method will not reoccur. Thus, a mixed method which integrates qualitative and quantitative methods is recommended because of its immense research contributions in terms of validation depth, and validation of best practices.

REFERENCES

Amusan, L. *et al.* 2016. Remodularising Technical Institutions Towards Manpower Delivery in Construction Sector in Nigeria, in *INTED Conference*. Valencia, Spain, 4991–4996.

Asaju, K. (2014). The Rising Rate of Unemployment in Nigeria: The Socio-Economic and Political Implications, *Global Business and Economic Research*, 3(2): 12–32.

Backa, L. & Wihersaari, M. 2014. Future engineering education: What competencies are energy companies looking for when recruiting graduates with a master of science (technology) degree? *Engineering Education*, 9(1): 2–17.

Badu-Nyarko, S. K. 2013. Quality assurance measures in distance learning at University of Ghana, *African Educational Research Journal*, 1(2): 126–133.

Bartlett, W. & Pagliarello, M. C. 2016. Agenda-setting for VET policy in the Western Balkans: employability versus social inclusion, *European Journal of Education*, 51(3): 305–319.

Boahin, P. & Hofman, A. 2013. A disciplinary perspective of competency-based training on the acquisition of employability skills, *Journal of Vocational Education & Training*. Routledge, 65(February 2015), 385–401.

Bowman, D. *et al.* 2016. Rusty, invisible and threatening: aging, capital, and employability, *Work, Employment & Society*, 95

Cardoso, Jose L. et. al. 2014. Employability and Higher Education, *Journal of Graduate Employability*, (0):17–31.

Chan, S. J., Lin, J. W. & To 2016. Aiming for better employment a holistic analysis from admission to labour market, *Journal of Higher Education Policy and Management*. Routledge, 38(3): 282–296.

Craft, E. L. *et al.* 2015. Aligning workforce skills with industry needs through problem-based learning environments, *ASEE Annual Conference and Exposition, Conference Proceedings*, 122nd ASEE(122nd ASEE Annual Conference and Exposition: Making Value for Society).

Creswell, J. W. 2013. Steps in Conducting a Scholarly Mixed Method Study, in *DBER Speaker Series*. (DBER SERIES).

Dacre Pool, L. & Sewell, P. 2007. The key to employability: developing a practical model of graduate employability, *Education + Training*, 49(4): 277–289.

Dania, J., Bakar, A. R. & Mohamed, S. 2014. Factors influencing the acquisition of employability skills by students of the selected technical secondary school in Malaysia, *International Education Studies*, 7(2): 117–124.

Davishahl, J. & Swisher, J. N. 2016. Advancing training pathways for the renewable energy workforce, *ASEE Annual Conference and Exposition, Conference Proceedings*, 2016–June.

Dean, S. A. 2017. ScholarWorks Soft Skills Needed for the 21st Century Workforce. Available at: http://scholarworks.waldenu.edu/dissertations.

Družovec, T. W. *et al.* 2013 Slovenian students on Projects/ Internships,*Proceedings of the 24th International Conference of the European Association for Education in Electrical and Information Engineering*, 210–214.

Gekara, V. O. & Snell, D. 2013. Employer responses to training and skills challenges in a market-based training system: the case of the transport and logistics industry Victor, *Journal of Chemical Information and Modeling*. Routledge, 53(9): 1689–1699.

Hillage, J. & Pollard, E. 1998. Employability : Developing A Framework For Policy Analysis Institute for Employment Studies, (85).

Ho, S. S. *et al.* 2016. Science Undergraduates Are Motivated to Undertake Leadership Education to Enhance Employability and Impact, 24(3): 71–83.

Husain, M. Y. *et al.* 2010. Importance of employability skills from employers' perspective, in *Procedia - Social and Behavioral Sciences*, 430–438.

Jackson, D. 2014. Testing a model of undergraduate competence in employability skills and its implications for stakeholders, *Journal of Education and Work*, 27(2): 220–242.

Jayaram, S. and Engmann, M. 2014. Developing skills for employability at the secondary level: Effective models for Asia, *Prospects*, 44(2): 221–233.

Khlaifat, A. & Qutob, H. 2013. Bridging the gap between oil and gas industry and academia, *Society of Petroleum Engineers - North Africa Technical Conference and Exhibition 2013, NATC 2013*, 2(4): 940–947.

Kumar, M., Garimella, U. & Nalla, D. 2014. Enabling higher order thinking and technical communication—An Indian context for OBE, *Frontiers in Education Conference (FIE), 2014*.

Lawal, A. W. 2013. Technical and Vocational Education, a Tool for National Development in Nigeria, *Mediterranean Journal of Social Science*, 4(8): 85–90.

Madar, A. R. & Buntat, Y. 2011. Elements of Employability Skills Among Students from Community Colleges Malaysia, 4(12): 1–11.

Mansour, B. El & Dean, J. C. 2016. Employability Skills as Perceived by Employers and University Faculty in the Fields of Human Resource Development (HRD) for Entry Level Graduate Jobs, *Journal of Human Resource and Sustainability Studies*, 4(3): 39–49.

Mazhar, N. & Arain, F. 2015. Leveraging on Work Integrated Learning to Enhance Sustainable Design Practices in the Construction Industry, *Procedia Engineering*. Elsevier B.V., 118, 434–441.

McDermott, R. *et al.* 2015. A comparative analysis of two globally distributed group projects: A perspective from CSCW/CSCL research, *Proceedings - Frontiers in Education Conference, FIE*, 2014.

McDermott, R., Bass, J. & Lalchandani, J. 2015. The learner experience of student-led international group project work in software engineering, *Proceedings - Frontiers in Education* Conference, *FIE*, 2015–February.

Mishra, M. 2014. Vertically integrated skill development and vocational training for socioeconomically marginalized youth: The experience at Gram Tarang and Centurion University, India, *Prospects*, 44(2): 297–316.

Patro, R. & Sahoo, H. K. 2016. Academic Embedded update and scope of employment, *Proceedings of the 2015 IEEE 3rd International Conference on MOOCs, Innovation and Technology in Education, MITE 2015*: 92–95.

Pieterse, V. & van Eekelen, M. 2016. Which Are Harder? Soft Skills or Hard Skills? BT - ICT Education: 45th Annual Conference of the Southern African Computer Lecturers Association, SACLA 2016, Cullinan, South Africa, July 5-6, 2016, Revised Selected Papers, in Gruner, S. (ed.). Cham: Springer International Publishing, 160–167.

Rahim Bakar, A. B., Mohamed, S. & Hamzah, R. 2013. An assessment of workplace skills acquired by students of vocational and technical education institutions, *International Education Studies*, 6(11): 15–20.
Renuga, M.& Ezhilan, S. 2014. Developing career skills of professional students through student mentees-alumni mentoring program, *Anthropologist*, 18(3): 705–716.
Saad, M. S. *et al.* 2013. Employers ' perception on engineering, information and communication technology (ICT) students employability skills, 15(1): 42–47.
Santos, P. R. dos *et al.* 2016. Inserção no mercado de trabalho e a empregabilidade de bacharéis em Biblioteconomia TT - Insertion in the labor market and the employability of graduates in Library, *Perspectivas em Ciência da Informação*, 21(2): 14–32.
Sgouropoulou, C. & Georgouli, K. 2014. Strengthening training to employment pathways through competency-based IT services, in *ACM International Conference Proceeding Series*.
Soler, J. & Andersson, P. H. 2014. Introducing process competences in a PBL-based engineering course, *2013 IEEE 5th International Conference on Engineering Education: Aligning Engineering Education with Industrial Needs for Nation Development, ICEED 2013*, 53–56.
Souto-Otero, M. & Shields, R. 2016. The investment model of volunteering in the EU-27 countries: volunteering, skills development, and employability. A multi-level analysis, *European Societies*, Euro (5): 487–513.
Tymon, A. 2013 The student perspective on employability, *Studies in Higher Education*, 38(6): 841–856.
Uddin, R. P. S. O. 2013. The Role of Technical and Vocational Education in Poverty Reduction among Youths in Nigeria, 4(4): 617.
Wedekind, V. & Mutereko, S. 2016. Higher education responsiveness through partnerships with industry: The case of a university of technology programme, *Development Southern Africa*. Taylor & Francis, 33(3): 376–389.
Zeti, K. & Faieza, A. A. 2014. Skills Improvement of Instructor to Provide Technical Consultation with Industries, *Advances in Mechanical and Manufacturing Engineering*, 564, 711–716.

TVET Towards Industrial Revolution 4.0– Hazirah Noh@Seth et al. (eds)
© 2020 Taylor & Francis Group, London, ISBN 978-0-367-24273-2

Exploring entrepreneurial competencies for technical college programs

Abubakar Ibrahim Muhammad, Yusri Bin Kamin & Nur Husna Binti Abd. Wahid
Department of Technical and Engineering Education, School of Education, Faculty of Social Sciences and Humanities, Universiti Teknologi Malaysia, Malaysia

ABSTRACT: One of the technical education challenges in Sub-Sahara Africa, including Nigeria, is the absence of an appropriate entrepreneurial competencies framework for the content of technical college curricula. The purpose of this study was to explore the entrepreneurial competencies required by technical college students for self-employment. The study was guided by an interview protocol and the data were quantitatively collected and analyzed. This research involved 15 participating experts who were purposively selected. The study found that managerial, marketing, financial, and opportunity-related entrepreneurial competencies, together with an entrepreneurial mindset, are worthwhile focuses for technical college programs in Sub-Sahara Africa and Nigeria, in particular. It was recommended that these entrepreneurial competencies should be integrated into technical college programs. Therefore, curricula should be reviewed to ensure that they meet education changes such as global trends, twenty-first century education strategies and Industrial Revolution 4.0, which has recently become widespread around the world.

1 INTRODUCTION

Technical and vocational education is an important key factor for national plans and industrial development worldwide; it is a tool to empower people, especially young people, for their sustainable and socioeconomic development (Bagale, 2015). The United Nations Educational and Scientific Organization (UNESCO) has defined technical and vocational education and training (TVET) as “those aspects of the educational process that involve, in addition to general education, the study of technologies and related sciences and the acquisition of practical skills, attitudes, understanding and knowledge related to employment in various sectors of economic life” (Companies Commission of Malaysia [CCM], 2013, p. 3).

It is clearly understood that, in developing countries, technical and vocational education is the most reliable instrument for easing poverty, unemployment and eliminating diseases and so on (Nwachukwu, 2014). The recognition of TVET as an important factor in economic growth has been maintained by UNESCO’s meeting of international experts on ‘Learning for work, citizenship and sustainability’ in that ‘Given that education is considered as a key to effective development strategies, TVET must be a fundamental key that can alleviate poverty, promote peace, conserve the environment, improve the quality of life for all and contribute to sustainable development’(UNESCO-UNEVOC, 2004, P. 2). In view of the foregoing, the Federal Republic of Nigeria (FRN) in the National Policy on Education (NPE), (Federal Republic of Nigeria, 2013:16) states that TVET covers three important areas which include technical colleges.

Surely, TVET focused on employment and self-reliance that is why it gives emphasis on practical competencies as well as skills acquisition in line with academic practices; therefore, the integration of TVET with entrepreneurial competencies alleviates most of the misery of the teaming young generation. UNESCO, (2012) in the Shanghai Consensus, recommended developing incentive frameworks and mechanisms to encourage active participation of stakeholders in planning, governance, curriculum, development and assessment of qualifications, as well as cooperation between the school enterprise and workplace learning.

Entrepreneurship education is a type of education that makes a person to become a responsible and enterprising person through the development of an entrepreneurial competence (European Commission [EC], 2017). This type of education helps people develop skills, knowledge and attitudes required to realize their goals. Research has shown that people with entrepreneurial education are more employable (EC, 2017). Therefore, the importance of entrepreneurial education for sustainable industrialization and poverty reduction cannot be underestimated. This phenomenon is essential in terms of creating qualified technical entrepreneurs, capable of encouraging investment opportunities, creating jobs and increasing productivity (Muhammad, 2015).Therefore, entrepreneurship can be defined as the performance of becoming an entrepreneur that can be stated to as someone who undertakes innovations, finances, business ideas in an attempt to transform skills and innovations into economic assets.

Nevertheless, the concept of entrepreneurship has an extensive meanings. The origin of the word entrepreneur is a French word "*Entreprendre"* means "taking" and in a business context means starting a business. Although entrepreneur can also be viewed from a variety of perspectives, the entrepreneur is a very high-minded person changing and pioneering, and represents features found in only a small fraction of the population. In another angle can see as a person who wants to work for himself. Gohain, Chakraborty and Saha, (2017) point out that entrepreneur is a person who can focus on the environment, identify opportunities and improve available resources and maximize opportunities. Kaur, (2015) maintained that entrepreneurship is a person's propensity to organize their business and manage profitability, with all the qualities of leadership, decision-making and management caliber, etc. The (International Center for the Technical and Vocational Education and Training [UNESCO-UNEVOC], 2016a, p. 6) has accepted the definition of Schoof, (2006)) "Entrepreneurship is the recognition of a possibility to create value, and the process act on this opportunity, whether it involves or not involves new entity. Whereas concepts such as "innovation" and "risk taking" in particular, are usually associated with entrepreneurship, they do not necessarily define the term'

In addition, the development of entrepreneurship, has been discussed and gathered more momentum over time, particularly during global economic crises. In the context of the crisis, at European level (Iacobuta and Socoliuc, 2014), Sam and van der Sijde, (2014) stated that the development of an entrepreneur is perceived as a main solution for the creation of jobs and sustainable economic growth. They also stated that, after these upheavals, numerous appeals and initiatives have been made to make entrepreneurship the engine of growth of the European economy and to establish the principles of 'think small first' at the center of national and European policies. Certainly, entrepreneurship is considered one of the key factors driving the forces of economic growth in Europe and in the world in general (Dvoulety, 2017). Ongoing discussions on entrepreneurship are based on different literature titles on business orientation and its measurement in various types and groups of companies (Codogni, Duda and Kusa, 2017). It has been discussed and tested in various conditions and in several geographic locations. However, according to Codogni, et al., (2017) academics identify a low number of publications related to the emerging economies of Central and Eastern Europe.

The promotion of entrepreneurship in the countries of Sub-Saharan Africa (SSA) has not played a vital role in the debate on education or entrepreneurship in the early years of independence (Kabongo, 2008). Organizations such as the African Development Bank, UNESCO and the World Bank have commissioned several studies to identify and investigate arrangements for integrating business education into technical institutions. Kabongo, (2008) reported shorter studies that addressed the issue of entrepreneurship education at graduate level and business programs offered at universities. Furthermore, he argued that most of the context of entrepreneurial research in SSA has focused on general principles, the theoretical mechanisms for creating entrepreneurs, and the implications of macroeconomic business success or failure. In summary, the role of entrepreneurial education in training and developing entrepreneurs and the enhancement of business activities in any economy cannot be underestimated.

Consequently, it can be clearly understood that it is necessary to integrate entrepreneurial competencies at technical colleges in Nigeria. Chekole, (2014) noted that lack of integration between the curriculum taught in schools and the skills required in the workplace in small businesses and enterprises is a major obstacle to the growth and development of Small-to-Medium Enterprises (SMEs). On this, Syed, (2013) and Enombo et al. (2015) recommended the need for a new school curriculum that embraces the entrepreneurial competencies, which serves as a solution to the current challenges of unemployment and high poverty. The intention of instilling the entrepreneurial mindset in students and producing new educated entrepreneurs and new businesses is to produce well-educated entrepreneurs who can create jobs (Ghina et al. 2015).

In spite of TVET's promising contribution to national and economic growth, the world is experiencing a worsening of the youth unemployment crisis. Youth unemployment, is more than 11 million in SSA at 2017, with the rate increasing rather than decreasing (International Labour Organization, 2016), which is a great challenge to all stake holders. Providing sound education, job opportunities and empowering youths though teaching entrepreneurial competencies helps youngsters to overcome the obstacles of their lives by exploring labor market and grow professionally (UNESCO-UNEVOC, 2016). Because entrepreneurial competency is an interdisciplinary training that emphasizes the tools necessary to start a new venture (Maigida et al. 2013) it is believed that this type of education is a key aspect in helping students become novice entrepreneurs (Lilleväli and Täks, 2017), so a systematic approach is needed to address the situation.

Consequently, it is necessary to know the characteristics and skills of entrepreneurship required to become a successful entrepreneur. Ismail et al. (2015) perceived that in order to achieve self-reflection on potential commercial competition, a person must study entrepreneurial competencies. Therefore it is highly important for developing countries such as Nigeria to embrace entrepreneurial competencies and integrate them at the technical college level. Competences have been defined (Kyndt and Baert, 2015) as combined and integrated components of knowledge, skills and attitudes. Thus entrepreneurial competencies (Rubin et al. 2017) have been identified as fundamental components for business success, hence they must be explored. Therefore, a successful entrepreneur needs a set of skills that leads to a sustainable and profitable venture.

Therefore technical colleges in Nigeria are in need of entrepreneurial competencies for self-employment upon graduation. Dawha and Medugu, (2016) believed that technical college student graduates without the required entrepreneurial competencies necessary for startup venture can be a menace to society. Similarly (Amaechi et al. 2017b) argued that technical college students lack entrepreneurial competencies essential for employment as qualified personnel. Therefore, it is obvious that technical college graduates suffer from many disadvantage upon graduation.

1.1 *Research objective*

To explore entrepreneurial competencies (ECs) required by technical college students for self-employment.

1.2 *Research question*

What are the ECs required by technical college students for self-employment?

2 RESEARCH METHODOLOGY

This study uses phenomenological research design, where data were collected through semi-structured interviews. This type of interview was utilized for the study because it offers interviewees the opportunity to present their views on the subject under consideration. The interview sessions involved 15 participants who are experts in their field of engagement. In fact, they were purposively selected to achieve the objective of the research (Bernard, 2002) and (Kumar, 2011) based on their knowledge and experience in the teaching and management of technical colleges as well as experience in entrepreneurship. To obtain saturated data, five were enterprise managers with more than 15 years of business experience, five were lecturers at technical institutions, three were administrators at technical institutions with managerial experience, and two participants worked in the Industrial Training Fund (ITF) as training officers. In consideration of the confidentiality assured to the interviewees, the researcher used a coding system of experts one (ER1) to 15 (ER15) to represent the interviewees.

The interview protocol consisted of two parts, the first part involved the demographics of the participants, while the second part was the main point of the interview. In the end, the data were transcribed and analyzed inductively using the thematic method. In the analysis of the qualitative content, the generated data were classified inductively (Creswell, 2014).

The in-depth interview was performed to explore ECs required by technical college students for self-employment in Nigeria. From the interview sessions, five main ECs were identified for technical college programs in Nigeria, including: Managerial Entrepreneurial Competencies (MECs), Marketing Entrepreneurial Competencies (MAECs), Financial Entrepreneurial Competencies (FECs) and Entrepreneurial Opportunity Competencies (EOCs) for technical colleges programs in Nigeria. Consequently, the study explores factors that influence the development of Entrepreneurial Mindset (EM) (which constitutes the fifth EC) of technical college students for self-employment. Furthermore, the purpose of the qualitative research was to determine the specific components of these constructs, to identify their dimensions or categories, and assess their relationships and how these competencies contribute to the improvement in start-up ventures through experts' views or perspectives. The outcome of this qualitative research answers research the question.

3 RESEARCH FINDINGS

The responses were typically identified according to the analysis of key ECs in this study. The codification process was developed by means of the three-Cs process *(Litchman, 2006)*. The experts' opinions were recorded during the interview sessions and the five key ECs featured in the interviews: managerial, marketing, financial, opportunity and EM. In general, the respondents have a similar characteristic of being experts in their fields and their answers appear to be similar, but with some differences. Therefore, in the in-depth interviews, to understand the important ECs which could be required by technical college students for self-employment in Nigeria, two main open-ended questions followed by probing questions were discussed with the interviewees. The experts were asked about their experience, skills, ideas, perspectives or views on concepts, components and important dimensions of ECs required by technical college students for self-employment. To this end, the main results of this study have been summarized as: ECs are important to technical college students because they lead them to become prospective entrepreneurs upon graduation, which helps them to become self-reliant and achieve self-employment. The respondents were asked to mention important ECs required by technical college students for self-employment.

To this end, the main results of this study have been summarized as: ECs are important to technical college students because it would leads them to become prospective entrepreneurs upon graduation which helps to become self-reliant/self-employment. The respondents were asked to mention important ECs required by technical college students for self-employment.

3.1 *Typical Responses by the Participants*

Some of the respondents' responses were as follows:

ER1: *"I believe managerial competencies are very important for technical college students since they are expected to establish their own business after graduation. So, they need all the leadership qualities. … ah secondly, I think they need to know about the market is important also likewise … ah … opportunities they have to look for market opportunities around them if they could try to seize the available market opportunities, they can easily become established. Another important competency is financial or funding because no business will be established without capital so if you have the available funds then one must know how to utilize those funds so you see here comes the need for financial competency. This is highly important as many businesses were established but because they lack financial discipline the business collapsed. Attitude change: - the reaction of one's thoughts, feeling and actions, it entails the adoption of new behavior pattern. It is obvious that with the attainment of the new behavior one may his needs. It is a known fact that when an individual's need is achieved, he develops new values and goals, this will invariably help him to advance. Entrepreneurship can be achieve through a compressive and well-formed attitude. Aha! Good that what I mean mindset … . We need to set their mind yes mindset."*

ER2: *"…hmm, what I would recommend as the important entrepreneurial competencies for technical college students to help them become prospective entrepreneurs are as follows, you see … 1. Financial discipline, this will guide them to make judicious use of the available funds or capital they have for the start of the business. No business without capital so they must have the ability of financial competency. 2. I will recommend managerial competency because I am anticipating these young people after graduation and establishing the business to be the leaders of that venture. So managerial competency is very important. 3. Knowledge of the market is important, more especially their community market, how it operates, what is the need of the market. 4. The available opportunity … is important to know how to snatch opportunity and how to develop that opportunity. Last but not least is how to make these young technicians accept this type of education, ah that … . Aha! Yes, entrepreneurial mindset is an important aspect of ECs because we need to inculcate entrepreneurial culture into them by setting their minds toward it".*

ER3: *"Hmmm … . Manager, ah I mean, managerial skills are needed for these students to become novice entrepreneurs because of administrative work, yes, they need it. Even before this, yes, since we are talking about technical college students, it is important to mention entrepreneurial mindset because we need first of all to set our youths' minds toward the importance of entrepreneurial activities. An entrepreneur requires financial skills; without financial skills the business will easily collapse, so it is important. Also, for a business to progress, one needs knowledge of the market that is the market segment at the place he operates. You also need to search for opportunity … yes, the available entrepreneurial opportunities are also very important."*

ER4: *"Hmmm, financial competency is very, very important in any business activity. Administrative ability … ah yes, managerial competency is important for these youngsters in preparing them to be self-employed. Also, they have to know how to seize business opportunities in trying to make the business successful. So, they have to look for a relevant market for their product. I think also we need, first of all, to even change their attitude toward entrepreneurial activities, we have to inculcate it into their mind … . Yes exactly, mindset, I mean."*

ER5: *"first of all for us to have ECs required by technical college students for self-employment is to instill or to set their mind toward entrepreneurial activities to enable them to start up their own business upon graduation that's entrepreneurial mindset. Then, they need to know how to manage the business when it was established; they need managerial competencies. Another important thing is the ability to control market; marking competency in this modern world is very important to every entrepreneur. Market opportunity should be taken into consideration because if you do not have the knowledge of how to search for an opportunity, it is very easy to lose the market segment close to you. Another one of the most important competencies is financial discipline because without the ability of financial record and control the business may surely die automatically."*

Going by the responses of the qualitative data, all the experts mentioned that managerial, marketing, financial and opportunity competencies are the most important ECs required by technical college students for self-employment. Likewise, the majority of the experts mentioned that an EM is an important component required by technical college students for self-employment.

The findings revealed the views of the experts on the attitude required for the development of technical college students' EM for self-employment in Nigeria. The majority of the experts agreed on the importance of EM to technical college students because, with a growth mindset, students could climb to the highest entrepreneurship level. This could be done by developing their attitudes toward entrepreneurship right from the college level. The participants expressed their views on how to develop students' EM. In his response, ER1 emphasized the need for nurturing entrepreneurship among technical college students:

> *"the reaction of one's thoughts, feelings and actions, it entails the adoption of new behavior pattern. It is obvious that with the attainment of the new behavior one may his needs. Entrepreneurship can be achieved through a comprehensive and well-formed attitude. Aha! Good that what I mean mindset . . .. We need to set their mind yes mindset."*

This is in agreement with Rekha *et al.* (2015) who found that nurturing EM through motivation and learning has led to the growth of entrepreneurship among Indians.

In their reaction experts (ER2, ER3 and ER5) stated that

> *". . ... Aha! Yes entrepreneurial mindset is an important aspect of ECs because we need to inculcate entrepreneurial culture to them by setting their minds toward it."*
>
> *". . . Even before this, yes since we are talking about technical college students it is important to mention entrepreneurial mindset because we need first of all to set our youth minds towards the importance of entrepreneurial activities."*
>
> *". . .. to instill or to set their mind towards entrepreneurial activities to enable them start up their own business upon graduation that entrepreneurial mindset."*

These findings are similar to those of Rae and Melton, (2016) who revealed that the introduction of entrepreneurial learning and mindset concepts developed by the KEEN (Kern Entrepreneurial Education Network) project across US engineering programs demonstrated reliable results among the students trained. Thus, it instilled an EM in the students. Therefore, from the explanation of the experts, it is clear that nurturing EM is one of the most important components required for the proposed EC model for technical college programs.

Based on the qualitative data, 14 (93%) of the experts agreed that *'managerial entrepreneurial competencies play a vital role in running entrepreneurial activities as fundamental sources to get profit and influence administration behavior'*. This is in agreement with (Rezaeizadeh et al. (2017) who affirmed that providing students with managerial competencies is one of the basic requirements of producing students with leadership abilities that will lead them to become prospective entrepreneurs. All experts unanimously stressed that managerial competencies should be given specific attention in an attempt to achieve long-term prosperity. This is in line with the findings of Ogbuanya and Nungse, (2017) who recommended the inclusion of MECs in the electronic curriculum of technical colleges in north-central Nigeria because the curriculum content was seriously inadequate for the development of MECs.

To sum up, the experts presented common perspectives and understandings of the importance of MECs for technical colleges.

Furthermore, the study revealed that MAECs are another important EC component required by technical college students for self-employment upon graduation. This is also in line with the findings of Ogbuanya and Nungse, (2017) who recommended the inclusion of MAECs in the electronic curriculum of technical colleges in north-central Nigeria because the curriculum content was seriously inadequate for the development of MAECs.

Generally, the interviewees believed that marketing is one of the vital ECs needed by entrepreneurs after having equipment and working capital. Some of the views of the experts are as follows:

ER5: *Another important thing is the ability to control the market; marketing competency in this modern world is very important to every entrepreneur.*
ER7: *. . .. marketing competencies will enable an entrepreneur to penetrate and explore the emerging market around him. . ..*
ER9: *To my own understanding, knowledge of marketing and the ability to assess the market are vital ECs needed in this study.*

These findings are in line with the observations of Chernysheva et al. (2017), who maintained the importance of developing MAECs in college students that should be manifest in their overall character. The expert responses continued:

ER11: *The important components for entrepreneurial training that are deemed necessary to be integrated in technical college should include . . . ability to market your product, yes, I mean marketing competencies; it has a paramount importance in this regard . . ..*
ER14: *. . . one of the important ECs needed for integrating into technical college programs is marketing skills, it is a very important aspect of entrepreneurship.*
ER15: *The knowledge of how to penetrate the market is a very important aspect of entrepreneurship.*

These findings were in agreement with the work of Ismail et al. (2015), who believed that MAECs are a significant component of the ECs needed by college students to become successful entrepreneurs. Therefore, MAECs should be among the important components of ECs required by technical college students for self-employment. Consequently, positive and continuous development of marketing strategies eventually affect and influence the entrepreneurial competence of technical college students as entrepreneurs and, consequently, improve their participation in a market segment.

The interview results also present FECs as an important component of ECs required by technical college students for self-employment upon graduation. Expert ER4 has this to say: '*. . . financial competency is very, very important in any business activity*', while ER2 emphasized that: '*Financial discipline will guide them to make judicious use of the available funds or capital they have for the start of the business; no business without capital, so they must have the ability of financial competency*'. According to ER3: '*An entrepreneur requires financial skills; without financial skills the business will easily collapse, so it is important*'. These findings were in agreement with Oztemel and Ozel, (2018), who believed that financial control protracted the venture life cycle.

In a nutshell, the experts unanimously emphasized the importance of FECs as one of the vital components of ECs for technical college students that could be integrated into the curriculum for an appropriate EC model.

Furthermore, the experts unanimously agreed that EOCs are also an important component of the ECs required by technical college students for self-employment in Nigeria, as revealed in the findings of this study.

Consequently, according to the most experts (ER1, ER2, ER3, ER5, ER6, ER7, ER9, ER10, ER11, ER12, ER13 and ER15) opportunity is one of the vital component of ECs in business. In their view, experts ER1, ER3, ER4 and ER5 indicated that: *"it is very important for an entrepreneur to always develop the business opportunity in accordance with the market needs. So when students are trained on how to develop their opportunities in trying to stat up venture creation it will lead them to a greater height."* This is in line with Sołoducho-Pelc,(2015) who argued that the development of entrepreneurial opportunity is dynamic and goes along with the current situations; as a result, the approach to entrepreneurial opportunity has to focus on attitude of the learner. It has been understood that opportunities begin as simple ideas that become more innovative as the entrepreneurs develop them. Consequently, the experts commonly expressed the view that opportunity recognition in business has paramount importance to entrepreneurs, with the majority of the experts pointing to opportunity recognition as an important component of EOCs. For instance, ER2, is a paint and foam producer and marketer for 22 years has this to say *"I am sure one of the most important competency is, teach the students how to recognize or seize opportunity; in essence this will bust the entrepreneurial mindset of students to become entrepreneurs after graduation. Because a person with skills in technical education was usually trained to produce products, so that he needs the knowledge of how to seize market opportunity so that he can attract people's interest in the consumption of his product."* This is in agreement with Sołoducho-Pelc, (2015) who opined that entrepreneurial opportunity development can be improved, that opportunities are dynamically developing and changing together with the changes that occur; therefore, in the approaching opportunities it is challenge to take an active attitude. It can be deduce that opportunity recognition gives a person the capacity to observe changed situations or overlooked possibilities in the environment that represent potential sources of profit or return to the venture as a great market strategy. Therefore, OECs is avital component worth for inclusion in the proposed ECs model

3.2 *Discussion*

From the data analyzed a new proposed ECs was established in order to develop non-technical skills for technical college students.

Therefore, as from the analyzed data it can be deduced that EM is the most important component of ECs that could first be instill in the mind of the students. Since EM is a forerunner of behaviors, intentions or business actions and represents the world view of an individual: their attitudes, dispositions, motivations and their expectations. It is likely that the mindset is influenced by non-cognitive and affective thinking. At the other extreme is the creation of new businesses, either in the form of a new business or within an existing organization (Rae and Melton, 2016). Therefore there is need in all the Sub Sahara Africa, Nigeria inclusive to take a fore front action in the integration of EM in the content of technical college programs for better nurturing of novice entrepreneurs that will go along side with the current 4.0IR development. Because this is one of the most important ways to position the perception of the youths to understand the importance of entrepreneurship and how to integrate multidisciplinary knowledge for the development of the entrepreneurial spirit to face the challenges of 4.0IR (Yeung, 2015). Therefore it hopes that this proposed ECs framework will be useful for the integration of EM into technical college programs for rapid development toward the realization of 4.0IR.

Consequently, managerial entrepreneurial competencies (MECs) was found to be an important component of ECs required by the students to establish new business. Since the students are expected to become leaders of business in the future. In fact MECs are specific type of individual competencies which help an individual to become an excellent manager, example of such are specific knowledge, abilities, skills, traits, motives, attitudes and values necessary to improve management performance (Fejfarová and Urbancová, 2015). This is supported by Chong, (2013) who affirmed that management performance based competencies are assessed through observed behaviours. It is imperative to include MECs among the ECs required by technical college students because it contribute significantly to career success and entrepreneurship activities. Ibrahim and Soufani, (2002) maintains that managerial competencies take the second set of factors related with successful entrepreneurs. Consequently, Brown and Hanlon, (2016) identifies MECs as an important aspect needed for early growth business.

It is imperative to include marketing entrepreneurial competencies (MAECs) in the technical college programs as opined unanimously by the experts. Due to the important of marketing in the ECs framework, Chernysheva, Kalygina and Zobov, (2017) reiterated that marketing is growing of both "inward" i.e. during the entire venturing system process, and "outwards" i.e. within the whole community socially and politically. Therefore, it is vital to note these two aspect of marketing evaluation in term of the students' marketing competencies. Since marketing tools and technologies emerge as complete and necessary elements, not only for the proper management of commercial enterprises, but also for politics, education, medicine, sports and non-profit organization (Chernysheva et al., 2017). Indeed, it is tremendously important to be included in technical college programs. A well organized and established marketing competencies for the training of technical students will help in boosting the nation economy, reduce crises, increase local/international donations, encouraging students' performance, etc. Consequently, it has been believed that student's desire to be an entrepreneur has great influence by their satisfaction with entrepreneurial marketing issues (Yousaf, Altaf, Rani and Alam, 2013). Therefore, for the students to become successful novice entrepreneurs they need MAECs.

Financial entrepreneurial competencies (FECs) is one of the most important competency required by technical college students for self-employment as recommended by overall experts who participated in the qualitative study in this work. FECs has receive much attention in recent times, developing and the developed countries have become increasingly concerned about the level of financial competency of entrepreneur. Dearth of financial competencies is one of the much challenges facing SMEs in the financial context. Eniola and Entebang, (2015). Consequently, technical college students are also liable of facing such issue when start up venturing (Jelilov and Onder, 2016) enumerated issue in financial competencies as to start a venture, an entrepreneur needs funds to start the venture and also to keep it buoyant. Therefor FECs is worthy to be mentioned for inclusion into technical college programs for producing successful prospective technical entrepreneurs.

Opportunity identification is the starting point of entrepreneurial process as it has been pointed by the experts in the qualitative study. Therefore, recognizing the initial phase is seen as a critical skill that students need during their training that will lead them to improve their capacity for innovation, proactivity and willingness to risk taking (Lindberg, Bohman, Hulten, and Wilson, 2017). It can be realized that the recognition of the opportunity gives a person the ability to observe the changed situations or the possibilities that are observed in the environment that represent potential sources of profit or return to the venture as a great market strategy.

Therefore, it is imperative to help students to develop attitudes that lead them to identify markets opportunities. Consequently, Sołoducho-Pelc, (2015) argued, the development of entrepreneurial opportunity is dynamic and goes along with the current situations, as a result the approach to entrepreneurial opportunity has to focus on attitude of the learner. Therefore, it is important to include this aspect in the proposed ECs model for the technical colleges.

Therefore, opportunity entrepreneurial competencies are important component worth for inclusion in the proposed ECs for technical college programs.

3.3 *Conclusions*

The main purpose of conducting the qualitative research in this study was to achieve research objective (to explore entrepreneurial competencies required by the technical college students for self-employment in Nigeria). Consequently, the study found that MECs, MAECs, FECs, EOCs and EM are all required by the technical college students for self-employment in Nigeria. Therefore, there is a need for all stakeholders in technical education in SSA and Nigeria, in particular, to use the recommendations of this study as they will enhance the much-needed desire for self-reliance and self-employment among African youth, in terms of poverty alleviation and sustainable peace, harmony and economic growth in the continent

3.4 *Recommendations*

Based on the findings of the study it has been recommended that these ECs (managerial, marketing, financial, entrepreneurial opportunity and EM) are worthy for technical college programs in SSA and Nigeria, in particular. Therefore, they should be integrated into technical college programs.

Consequently, the curriculum should be reviewed to meet with the education changes such as global trends, twenty-first century education strategy or Industrial Revolution 4.0, which have recently spread all over the world.

REFERENCES

Amaechi, O. J., Orlu, I., Obed, O. O. and Thomas, C. G. (2017). The role of technical and vocational education and training (TVET) as agents of tackling unemployment and poverty amongst technical college graduates in Rivers state. *World Journal of Engineering Research and Technology*, *3*(2), 17–30. Retrieved from www.wjert.orgSJIF

Bagale, S. (2015). Technical education and vocational training for sustainable development. *Journal of Training and Development*, *1*(1), 16–20. https://doi.org/10.3126/jtd.v1i0.13085

Bernard, H. R. (2002). Research methods in anthropology: Qualitative and quantitative approaches.

Brown, T. C. and Hanlon, D. (2016). Behavioral criteria for grounding entrepreneurship education and training programs: A validation study. *Journal of Small Business Management*, *54*(2), 399–419. https://doi.org/10.1111/jsbm.12141

Chekole, Z. G. (2014). *Challenges and prospects of micro and small enterprises in Awi zone: The case of Dangila district.* Indra Gandhi National Open University. Retrieved from http://www.googlescholar.com

Chernysheva, A., Kalygina, V. and Zobov, A. (2017). Marketing and PR activities of the leading world universities: modern tools and development trends. In D. Vrontis, Y. Weber, and E. Tsoukatos (Eds.), *10th Annual Conference of the EuroMed Academy of Business Global and national business theories and practice: bridging the past with the future* (pp. 347–357). Rome: EuroMed.

Chong, E. (2013). Managerial competencies and career advancement: A comparative study of managers in two countries. *Journal of Business Research*, *66*(3), 345–353. https://doi.org/doi:10.1016/j. jbusres

Codogni, M., Duda, J. and Kusa, R. (2017). Entrepreneurial orientation in high-tech and low-tech SMEs in Malopolska region, *18*(1), 7–22. Retrieved from http://dx.doi.org/10.7494/manage.

Companies Commission of Malaysia. (2013). Education for Sustainable Development : Promoting Technical Education and Vocational Training. https://doi.org/10.1093/acprof:oso/9780195171648.003.0022

Dawha, J. M. and Medugu, J. D. (2016). Emerging entrepreneurial and business planning competencies required by motor vehicle mechanic students in establishing entreprise in Bauchi and Gombe states, Nigeria. *International Journal of Humanities Social Sciences and Education (IJHSSE)*, *3*(1), 156–161. Retrieved from htt//www.googlescholar.com

Dvoulety, O. (2017). What is the Relationship between Entrepreneurship and Unemployment in Visegrad Countries? *Central European Business Review*, *6*(2), 42–53.

Eniola, A. A. and Entebang, H. (2015). Financial literacy and SME firm performance. *International Journal of Research Studies in Management*, *5*(1). https://doi.org/10.5861/ijrsm.2015.1304

Enombo, J. P., Hassan, S. L. and Iwu, C. G. (2015). The significance of entrepreneurship education in Gabonese schools: justifications for a new curriculum design. *Socioeconomica – The Scientific Journal for Theory and Practice of Socio-Economic Development* *4*(8), 493–506. https://doi.org/dx.doi.org/10.12803/SJSECO.48139 JEL:

European Commission. (2017). European commission. Retrieved November 19, 2017, from https://ec.europa.eu/jrc/en/publication/eur-scientific-and-technical-research-reports/entrecomp-entrepreneurship-competence-framework

Federal Republic of Nigeria. (2013). National policy on education. Lagos: Nigerian Educational Reseach and Development Council. Retrieved from http://www.google.com

Fejfarová, Martina and Urbancová, H. (2015). Application of the competency-based approach in organisations in the Czech Republic. *Ekonomika a Management*, *18*(1), 111–122. https://doi.org/10.15240/tul/001/2015-1-009

Ghina, A., Simatupang, T. M. and Gustomo, A. (2015). Building a systematic framework for entrepreneurship education. *Journal of Entrepreneurship Education*, *18*(2), 73–99.

Gohain, D., Chakraborty, T. and Saha, R. (2017). Are entrepreneurs trainable towards success : Reviewing impact of training on entrepreneurship success are entrepreneurs trainable towards success : Reviewing impact of training on entrepreneurship success. *Training & Development Journal*, *8*(1), 44–59. https://doi.org/10.5958/2231-069X.2017.00006.3

Iacobuta, A-O. and Socoliuc, O.-R. (2014). *European entrepreneurship in times of crisis : realities, challenges and perspectives* (No. L26; M13; Y1) (Vol. VI). Alexandru. Retrieved from www.google.com

Ibrahim, A. B., and Soufani, K. (2002). Entrepreneurship education and training in Canada: a critical assessment. *Education + Training*, *44*(8/9), 421–430. https://doi.org/10.1108/00400910210449268

International Labour Organization. (2016). *World employment social outlook.* Geneva. Retrieved from http://www.ilo.org/wcmsp5/groups/public/—dgreports/—dcomm/—publ/documents/publication/wcms_513739.pdf

Ismail, V. Y., Zain, E. and Zuliha, (2015). The portrait of entrepreneurial competence on student entrepreneurs. *Procedia - Social and Behavioral Sciences*, *169* (August2014), 178–188. https://doi.org/10.1016/j.sbspro.2015.01.300

Jelilov, G. and Onder, E. (2016). Entrepreneurship : Issues and solutions evidence from Nigeria. *Pyrex Journal of Business and Finace Mnagaement Research*, *2*(3), 10–13. Retrieved from https://scholar.google.com/scholar
Kabongo, J. D. (2008). *The Status of entrepreneurship education in colleges and universitis in Sub-Saharan Africa*. Retrieved from https://www.researchgate.net/profile/Jean_Kabongo2/publication/228466849_The_status_of_entrepreneurship_education_in_colleges_and_universities_in_sub-Saharan_Africa/links/00b7d537e5f985c32f000000.pdf
Kaur, G. (2015). Women entrepreneurship motivated by pull and push factors. *Human Rights International Research Journal*, *3*(1), 281–284. Retrieved from http://www.googlescholar.com
Kumar, R. (2011). *Research methodology a step-by-step guide for beginners*. (R. Kumar, Ed.), *Mixed Souces FSC* (3rd ed.). Los Angeles: Sage publications Ltd. https://doi.org/http://196.29.172.66:8080/jspui/bitstream/123456789/2574/1/Research%20Methodology.pdf
Kyndt, E. and Baert, H. (2015). Entrepreneurial competencies: Assessment and predictive value for entrepreneurship. *Journal of Vocational Behavior*, *90*, 13–25. https://doi.org/10.1016/j.jvb.2015.07.002
Lilleväli, U. and Täks, M. (2017). Competence models as a tool for conceptualizing the systematic process of entrepreneurship competence development. *Education Research International*, *2017*, 1–16. https://doi.org/10.1155/2017/5160863
Lindberg, E., Bohman, H., Hulten, P. and Wilson, T. (2017). Enhancing students' entrepreneurial mindset: a Swedish experience. *Education + Training*, *59*(7/8), 768–779. https://doi.org/10.1108/ET-09-2016-0140
Litchman, M. (2006). Qualitative research in education: A user's guide. London: Sage publications. Retrieved from https://books.google.com.my/books/about/Qualitative_Research_in_Education.html?id=_q0xZ93fLGwC&redir_esc=y
Maigida, J. F., Saba, T. M. and Namkere, J. U. (2013). Entrepreneurial skills in technical vocational education and training as a strategic approach for achieving youth empowerment in Nigeria. *International Journal of Humanities and Social Science*, *3*(5), 303–310. Retrieved from www.ijhssnet.com
Muhammad, A. I. (2015). Lifelong training in TVET for sustainable industrial development in Nigeria. *Journal of Nigerian Association of Teachers of Technology (JONATT)*, *10*(3).
Nwachukwu, O. P. (2014). Poverty reduction through technical and vocational education and training (TVET) in Nigeria. *Developing Country Studies*, *4*(14), 10–14. Retrieved from http://www.iiste.org/Journals/index.php/DCS/article/viewFile/14053/14361
Ogbuanya, T. C. and Nungse, N. I. (2017). Adequacy of electronics curriculum in technical colleges in North Central Nigeria for equipping students with entrepreneurial skills. *Industrial Engineering Letters*, *7*(5), 1–6. Retrieved from www.iiste.org
Oztemel, Ercan and Ozel, S. (2018). Financial Competency Assessment Model. *Journal of Business & Financial Affairs*, *07*(01), 1–7. https://doi.org/10.4172/2167-0234.1000317
Rae, D., and Melton, D. E. (2016). Developing an entrepreneurial Mindset in US engineering education : An international view of the KEEN project. *Journal of Engineering Entrepreneurship*, *7*(3), 1–16.
Rekha, S. K., Ramesh, S. and Jaya Bharathi, S. (2015). Empherical study on the relationship between entrepreneurial mindset and the factors affecting intrapreneurship: A study in Indian context. *International Journal of Entrepreneurship*, *19*, 53–59.
Rezaeizadeh, M., Hogan, M., O'Reilly, J., Cunningham, J. and Murphy, E. (2017). Core entrepreneurial competencies and their interdependencies : insights from a study of Irish and Iranian entrepreneurs, university students and academics. *Int Entrep Manag J*, *13*, 35–73. https://doi.org/10.1007/s11365-016-0390-y
Rubin, Y., Lednev, M. and Mozhzhukhin, D. (2017). A competency-based approach to bachelor's degree entrepreneurship erograms. In *United States Association for Small Business and Entrepreneurship. Conference Proceedings* (pp. 341–376). Boca Raton: ProQuest. Retrieved from https://search.proquest.com/openview/aa1efc64c904af174fd13b3c5d084fda/1?cbl=38818&pq-origsite=gscholar
Sam, C. and van der Sijde, P. (2014). Understanding the concept of the entrepreneurial university from the perspective of higher education models. *Higher Education*, *68*(6), 891–908. https://doi.org/10.1007/s10734-014-9750-0
Schoof, U. (2006). *Stimulating Youth Entrepreeurship: Barriers and incentives to enterprise start-ups by young people* (on Youth and Entrepreneurship SEED No. 76). *Ilo*. Geneva. Retrieved from http://www.ilo.org/youthmakingithappen/Resources/01.html%5Cnpapers2://publication/uuid/E0F5D26F-9738-4EBE-A831-715E7EBBBE90
Sołoducho-Pelc, L. (2015). Searching for opportunities for development and innovations in the strategic management process. *Procedia - Social and Behavioral Sciences*, *210*, 77–86. https://doi.org/10.1016/j.sbspro.2015.11.331

Syed, Z. A. (2013). The need for inclusion of entrepreneurship education in Malaysia lower and higher learning institutions. *Int Entrep Manag J*, *55*(2), 191–203. https://doi.org/10.1007/s11365-016-0390-y
UNESCO-UNEVOC. (2004). TVET FOR SUSTAINABLE DEVELOPMENT - OPPORTUNITIES AND CHALLENGES -. Retrieved November 29, 2017, from http://www.unevoc.unesco.org/filead min/user_upload/docs/Vietnam06_backgr_paper.pdf
UNESCO-UNEVOC. (2016). Making youth entrepreneurship a viable path. How can TVET institutions help promote entrepreneurship. Retrieved October 21, 2017, from www.unevoc.unesco.org
UNESCO. (2012). Shanghai Consensus: Recommendations of the Third International Congress on Technical and Vocational Education and Training 'Transforming TVET: Building skills for work and life.' Retrieved December 20, 2017, from http://www.unesco.org/fileadmin/MULTIMEDIA/HQ/ED/pdf/outcomesdocumentFinalwithlogo.pdf
Yeung, S. M. C. (2015). A mindset of entrepreneurship for sustainability. *Corporate Ownership and Control*, *13*(1CONT7), 797–811.
Yousaf, U., Altaf, M., Rani, Z. and Alam, M. (2013). Improving entrepreneural marketing learning: A study of business graduates, Pakistan. *Journal of Women's Entrepreneurship and Education*, *1*(2), 74–89.

TVET Towards Industrial Revolution 4.0– Hazirah Noh@Seth et al. (eds)
© 2020 Taylor & Francis Group, London, ISBN 978-0-367-24273-2

Approaches of integrating sustainability in to higher education curricula: A review

N. Mukhtar, M.S. Saud & Kamin
Department of Technical and Engineering Education, Faculty of Social Sciences and Humanities, Universiti Teknologi Malaysia, Johor Bahru, Malaysia

ABSTRACT: The importance of integrating sustainability in to the higher education curricula is crucial as affirmed by charters, declarations, and UN's proclamation of 2005—2014 as the decade for education for sustainable development. Therefore, researchers employ diverse approaches of sustainability integration for higher education programmes. This study investigated the numerous approaches in use, and identified the leading and predominant approach. The authors conducted a systematic review of literature with the analysis of 30 articles from various disciplines. The researchers classified the approaches in to vertical and horizontal integrations. Vertical integration includes modular/bolt-on, PBL, PoBL, role playing games, online, and iterative learning. While horizontal integrations are PPBL, systems approach, integrative learning, dialectic approach, and university partnership. Modular/Bolt-on and PPBL approaches emerged as the leading and predominant approaches. The researchers hoped the findings will be useful to higher education institutions especially in sub-Sahara Africa for integrating sustainability in to their programmes.

1 INTRODUCTION

From time immemorial, human unsustainable thinking was what exposed the world to the various environmental, economic, and social catastrophes (Doppelt, 2012). Only after people begin to think sustainably, will the technologies and policies of sustainable development become realizable; and education as a key can greatly help in this direction. The United Nations identified Education and particularly Higher Education Institutions (HEIs) as the most convenient and appropriate vehicle for advancing training and awareness for sustainability, and thus achieve the much desired goals of sustainable development. This was the main reason United Nations' declared 2005—2014 as the decade for education for sustainable development (DESD) in Johannesburg (UNESCO, 2002). Although, even before the proclamation some institutions have in one point of time attempted to incorporate sustainability in their different educational aspects including the curriculum (Segalas, et al, 2010), the declaration encouraged most Higher Education Institutions to quicken the pace towards dealing with the pragmatic issues related to integrating sustainability in their diverse activities (Dubois et al, 2010).

Carroll, (2005) summarized the work of embedding sustainable development into the curriculum that includes: workshops with academic staff to raise awareness; analysis of courses to identify those that already include, and/or could readily incorporate sustainable development principles and teachings; identifying needs of academic teaching staff for sustainable development material; identification of synergies and collaboration for sustainable development with other disciplines; and an introductory sustainable development lecture for all first year students. Embedding sustainability in to education system plays a vital role towards the realization of sustainable development goals. However, in many under-developed countries, especially in Sub Sahara Africa (SSA), the issue of integrating sustainability in to curricula is still at far with the system of education. Therefore, a systematic study is timely to explore various approaches of sustainability integration in HEIs curricula across the globe, in order to identify and recommend a dominant approach to HEIs in West African sub-region.

Several sustainability educators and researchers in higher education across the globe, used and/or proposed models of the approaches for integrating sustainability in higher education curriculum. Thomas, Kyle, and Alvarez (1999), for instance identified a number of models and approaches with their concomitant methods and requirements such as Modular approach, Intra-disciplinary framework, Inter-disciplinary framework, Exploring course culture, Professional practices, Experiential learning, and Flexible learning resources. Other researchers in sustainability education identified other approaches such as Incremental approach (Lozano, 2016); Broad and General Approach (Sammalisto and Lindhquist, 2008); General Matrix approach (Rusinko, 2010); Bolt-on, Built-in and Re-built approaches (Huntzinger, 2007); and Value Integration session approach (Phiri, 2010). Similarly, Lidgren, Rodhe and Huising, (2006), and Scott and Gough (2006) described an approach that requires the need to think strategically about integrating sustainability in Higher Education (HE). Also Dubois et al, (2010) discussed on other approaches that include problem solving, projects, demonstration, role-playing, role-modeling, study tours, industrial placement, ICT tutorials and conventional approaches (lectures, seminars and workshops).

From the foregoing, many success stories and progress are recorded in the literature upon embedding sustainability in the HEIs curricula. However, despite the enormous literatures discussing various approaches, there have been little studies that identify the dominant approaches of sustainability integration in the global context. For instance, Thürer, Tomasevic, Stevenson, Qu and Huisingh (2017) limited their search to only one database (i.e., Scopus), and case-studies in engineering curricula; whereas this study stretched to other three bibliographic databases, as well as captures other discipline's curricula beyond engineering.

2 RESEARCH QUESTIONS

This is a systematic review that intends to study and synthesize the current literature related to the differing approaches of integrating sustainability in to HEIs curricula, with a sole aim of exploring and identifying the dominant approach for promoting the integration of sustainability in Africa south of Sahara. This paper obviously aimed to answer the following research questions:

1. What are the approaches of integrating sustainability in Higher Education Institutes (HEIs) curricula?
2. Which is the leading and predominant approach of integrating sustainability in Higher Education Institutes (HEIs) curricula?

3 METHODOLOGY

This paper adopts a systematic review of the literature to ascertain the relevant articles for inclusion in the study. As a result, this section described the sources, screening and analysis of the articles. The focal point of this paper is on the approaches/techniques of integrating sustainability in higher education institutions' curricula. Therefore, the authors searched four (4) bibliographic databases which include Scopus, Web of Science, ERIC and Science Direct using keywords: Sustainable AND Education, Education AND Sustainable AND Development, Higher AND Education AND Curriculum. This resulted in to a vast number of materials which makes it almost impossible to work on such large literatures. As such, to reduce the number of articles to a manageable size, the following criteria were set:

i. The authors restricted document type to 'articles'
ii. The authors considered only peer-reviewed articles published in 2014 to 2018 (period of 5 years); and
iii. The authors restricted the search to the title of the papers.

Apart from the above criteria, the authors did not set a boundary on the subject area and journal type. Consequently, the search produced the following articles as shown in the Table 1 below:

Table 1. Results of Retrieved Articles for each Keyword against the Database

Databases	Sustainable AND Education	Education AND Sustainable AND Development	Higher AND Education AND Curriculum	Total
Scopus	119	78	127	324
Web of Science	255	80	204	539
ERIC	220	158	403	781
Science Direct	269	613	154	1036
Total	863	929	888	2680

The Table 1 above shows that the use of the keywords Sustainable AND Education, Education AND Sustainable AND Development, Higher AND Education AND Curriculum across the four data bases produced 863, 929, and 888 articles respectively, which makes 2680 articles.

The 2680 articles that stand as the original sample reduced to 2010 because of duplicates removal through reading and comparing the titles and abstracts of the articles belonging to the three (3) categories based on the keywords. This number of articles (i.e., 2010) was further reduced to 588 articles by removing irrelevant articles. Irrelevant articles are those not concerned with sustainability in higher education. It was also observed that sustainability integration at higher education level is either at undergraduate or post-graduate levels. Bearing in mind the primary objective of this study, the authors excluded articles related with sustainability integration at post-graduate level, and the sample of 588 articles came down to 324. The authors further excluded all articles that focus on other aspects of university practices (and not exactly the curriculum) such as research outreach, campus operations, as well as assessment and reporting, bringing the number of sample to 107. This is because embedding sustainability is not limited to integration at the aspect of curriculum only, but that it includes also extra-curricular activities or what Winter and Cotton, (2012) described as 'hidden curriculum'. Additionally, out of this number, some papers did not explicitly state the approach or techniques they employed in integrating the sustainability in to the curriculum. We therefore, read these papers once again trying to envision which approach did they exactly use but to no avail, and were later removed from the study which reduced the sample further to 56. The authors could only retrieved 30 articles using various methods or channels of information retrieval, and form the final sample for the study. The final sample of the articles was subsequently analyzed qualitatively using content analysis and summarized as shown in the Table 2 below.

Table 2. Analysis of Approaches Used for Integrating Sustainability in Higher Education Curriculum

Studies	Journal	Study Foci/Design	Approaches
Shing, Mohamad & Raman, (2016)	Int'l Journal of Engr. Education, Vol:32, Issue:6	Survey	Bolt-on approach
Iyer-Raniga, & Andamon,. (2016).	Int'l Journal of Sustainability in Higher Education, Vol:17, Issue:1	Conceptual/View Point	Outcome/practice-led approach
Halbe, Adamowski &Pahl-Wostl, (2015)	Journal of Cleaner Production, Vol:106	Case-study	Iterative learning approach

(*Continued*)

Table 2. (*Continued*)

Studies	Journal	Study Foci/Design	Approaches
Du, Su, &Liu (2013)	Journal of Cleaner Production, Vol:61	Mixed-Method Research	Problem and Project based learning approach
Mintz & Tal, (2013)	Journal of Biological Education, Vol:47,Issue:3	Mixed-Method Research	System approach
Schmidt, Palekhov, Shvetz & Palekhova, (2016)	Naukovyi Visnyk Natsionalnoho Hirnychoho Universytetu Issue 2	Results of an On-going Project	University partnership
Al-Hagla, (2012)	International Journal of Architectural Research - Volume 6 - Issue 1 - March 2012	Mixed-Method Research	Micro and Macro analysis approach
Siddiqui, A/rasheed, Mohammed, Alsaikhan & Alhashem (2012)	ASEE Annual Conference and Exposition, Conference Proceedings	Stand-Alone course	Project Based Learning approach
Garland, Khan & Parkinson (2011)	Proceedings of E and PDE International Conference on Engineering and Product Design Education	Stand-Alone course	Problem Based Learning & Peer Assisted Learning
Guerra, (2017)	Int'l Journal of Sustainability in Higher Education, Vol:18, Issue:3	Qualitative research design	Problem Based Learning
Bjornberg, Skogh &Stromberg, (2015)	Int'l Journal of Sustainability in Higher Education, Vol:16, Issue:5	Qualitative research design	Study visits and role plays
Husgafvel, Martikka, Egas Ribiero & Dahl, (2017)	Int'l Journal of Higher Education, Vol:6, Issue:2	Stand-Alone course	Modular approach
Weatherton, Sattler, Mattingly, Chen, Rogers & Dennis (2012)	Forum of public policy, Online, v2012 n2 2012	Capstone project/ Descriptive	Modular approach
Biswas, (2012)	Int'l Journal of Sustainability in Higher Education, Vol:13, Issue:2	Conceptual/View Point	Bolt-on approach
Minano, Aller, Anguera & Portillo, (2015)	Journal of Technology and Science Education, Vol:4, Issue:4	Descriptive	Bolt-on approach
Khalifa & Sandholz, (2012)	Int'l Journal of Environmental and Science Education, Vol:7, Issue:2	Descriptive	Trans-disciplinary approach
McConville, Rauch, Helgegren & Kain, (2017)	Int'l Journal of Sustainability in Higher Education, Vol:18, Issue:4	Survey/Course evaluation	Role-Playing games approach
Hayden, Rizzo, Dewoolker, Neumann, & Sadek, (2011)	Advances in Engineering Education, Vol:2, Issue:4	Mixed-Method Research	System approach
Rhee, Oyamot, Parent, Speer, Basu & Gerston, (2014)	Advances in Engineering Education, Vol:4, Issue:2	Mixed-Method Research	Multidisciplinary Project-based Instruction approach
Nazzal, Zabinski, Hugar, Reinhart, Karwoski & Medani, (2015)	Advances in Engineering Education, Vol:4, Issue:4	Mixed-Method Research	Modular approach
Willamo et al., (2018)	Ecological Modelling, 370, 1–13	Case Study Descriptive	Dialectic approach

(*Continued*)

Table 2. (*Continued*)

Studies	Journal	Study Foci/Design	Approaches
Ortega-Sánchez et al., (2018)	Journal of Cleaner Production, 171, 733–742		Integrative learning approach
Anand, Bisaillon, Webster, & Amor, (2015)	Journal of Cleaner Production, 108, 916–923.	Descriptive	Partnership, Learning, Collaboration and Transfer in SDE (PACTE 2D)
Lozano & Lozano, (2014)	Journal of Cleaner Production, 64, 136–146	Curriculum Development & Evaluation	Integrative approach
Tejedor, Segalàs, & Rosas-Casals, (2018)	Journal of Cleaner Production, 175, 29–37	Qualitative Exploratory design	Trans-disciplinary approach
Rose, Ryan, & Desha, (2015)	Journal of Cleaner Production, 106, 229–238	Case Study	Web Portal approach
Annan-Diab & Molinari, (2017)	International Journal of Management Education, 15(2), 73–83.	Case Study	Interdisciplinary approach
Pérez-Foguet et al., (2017)	Journal of Cleaner Production, 172, 4286–4302	Mixed-Method Research	Online Training Course approach
Von Blottnitz, Case, & Fraser, (2015)	Journal of Cleaner Production, 106, 300–307	Curriculum Reformation	Problem and Project Base learning
Leal Filho, Shiel, & Paço, (2016)	Journal of Cleaner Production, 133, 126–135	Qualitative-Multiple Method Design	Project-Based Learning

4 RESULTS

Research Question 1: What are the approaches of integrating sustainability in Higher Education Institutes (HEIs) curricula?

Table 2 analyzed the approaches of integrating sustainability in HEIs with specific reference to undergraduate curricula. The table reveals the diverse approaches researchers employed to integrate sustainability in to the curricula. These are the Modular approach, Multidisciplinary approach, System approach, Role-Playing games approach, Trans-disciplinary approach, Bolt-on approach, Study visits and role plays, Problem Based Learning, Problem Based Learning and Peer Assisted Learning, Project Based Learning approach, Micro and Macro analysis approach, Problem and Project Based Learning approach, Iterative learning approach, Outcome/Practice-led approach, Dialectic approach, Integrative approach, Online learning approach, and Inter-disciplinary approach.

The researchers observed with interest that some approaches share the same principles and processes with others and what just differentiated them is the nomenclature; as such the authors categorized them in to the following groups (with their frequencies) as shown below in Figure 1.

Approaches categorized as others include: Outcome/practice-led approach, Iterative learning approach, Micro and Macro analysis approach, and Dialectic approach.

A further analysis of Table 2 shows that the researchers also employed diverse study foci/designs in reporting their findings, with mixed-method research design as the most populous foci used by the scholars (4, 5, 7, 18, 19, 20, and 28); followed by Case-study design (3, 21, 26, and 27); Qualitative research design (10, 11, 25, and 30) and Descriptive design (15, 16, 22, and 23) with four appearances each. Other designs used are Stand-Alone course (8, 9, 12); Conceptual view points (2, 14); and survey (1, 17). The authors grouped designs mentioned only once under 'Others' category, and they include (6, 13, 24 and 29).

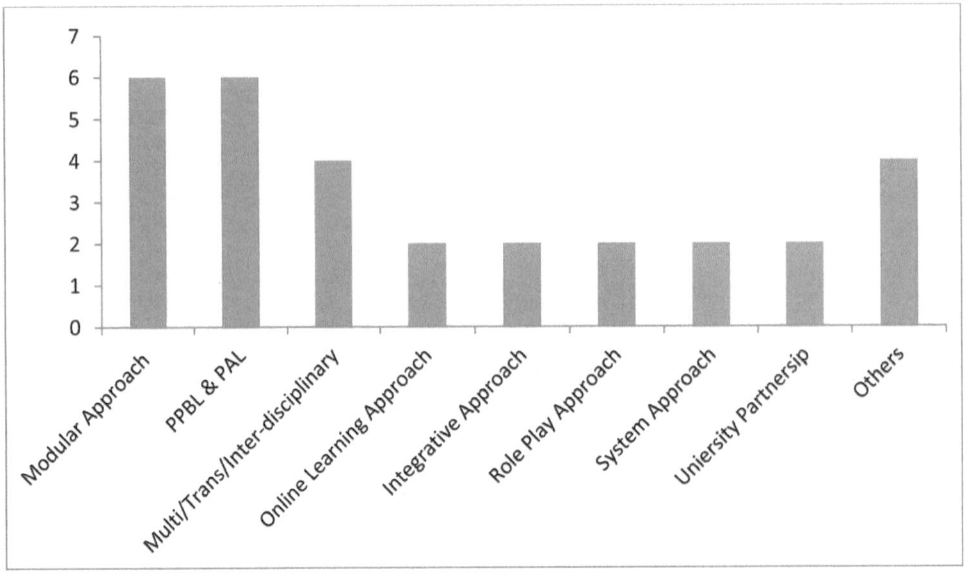

Figure 1. Frequencies of Approaches of Integrating Sustainability in HEIs Curricula

Research Question 2: Which is the leading and predominant approach of integrating sustainability in Higher Education Institutes (HEIs) curricula?

Table 2 further discloses that the leading and predominant approaches used by scholars and researchers are the Modular/Bolt-on approach and Problem and Project Based Learning (PPBL) approach. Six (6) studies each among the articles reviewed in this paper used these two approaches. Other frequent used approaches comprises Multi/Trans/Inter-disciplinary approach used in four (4) studies; Online Learning, University partnership, System approach and Role Playing Games used in two (2) studies each; while Outcome/practice-led approach, Iterative learning approach, Micro and Macro analysis approach, and Dialectic approach classified as 'others' were used in one (1) study each. Therefore, based on this analysis Modular/Bolt-on and PPBL are the leading and predominant approaches.

5 DISCUSSION

The chief purpose of this work was to investigate the approaches that researchers employed to integrate sustainability in the curricula of higher education, as well as to identify from the approaches, the leading and predominant approach used regarding undergraduate curricula. The numbers of studies involved are few because the authors restricted themselves to considering only a five-year period, which might subject the findings to lesser precision.

Thurer, at al, (2017) in their study on systematic review of literature on integrating sustainability in engineering curricula cited Kamp (2006), who presented three broad approaches of integrating SD in to the curriculum, which include embedding the concept of sustainable development into regular disciplinary courses; the design of a new elementary course; and, providing the option to graduate in a sustainable development specialization. Following Kamp's presentation of the approaches of integrating sustainability, Huntzinger, et al (2007) also analyzed models of sustainability integration and came up with three expansive approaches. These are: 'Re-built' or 'Re-design' approach, signifying wholly integration; 'Build-in' approach, which shows significant curricula changes; and 'Bolt-on' approach, implying cosmetic reform or just an awareness about SD. Cheah, Yang and Sale (2012), throw more light on the approaches suggested by Huntzinger, and refer the approaches using different terminologies as Transformation—i.e., Education as sustainability or 're-build' strategy; Reformation—i.e., Education for sustainability or 'built-in' strategy; and Bolting-on—i.e., Education about sustainability or 'add-on' strategy. They lamented that the peak form of integration is the Transformation or 'Re-build' approach, which can be described as inculcating the sustainability literacy in our graduates through understanding the immediate and long-term future of the ecology and socio-economy of societies and how they are impacted by human unsustainable actions; and through cultivating essential knowledge and skills that will make us adjust to a more sustainable way of doing things (Martin, 2008).

However, Watson, et al, (2013), lately compiled the diverse approaches of sustainability integration into two (2) broad categories, namely, vertical and horizontal integration. Watson described vertical integration as a specific course on SD added to the curriculum; whereas horizontal integration involves integrating sustainability spanning from, providing some coverage on SD issues in an existing course, intertwining sustainability in existing courses, to offering a sustainability specialization within an existing program or designing a specialist sustainability degree. Watson's vertical integration corresponds to Huntzinger's Bolt-on approach, Cheah's add-on strategy and Kamp's design of a new elementary course. Similarly, the horizontal integration described by Watson conforms to 'Re-built' and 'Build-in' approaches; Transformation and Reformation approaches; and embedding the concept of SD into regular disciplinary courses and option to graduate in a SD specialization approaches, described by Huntzinger, Cheah and Kamp respectively. This study therefore, employed the vertical and horizontal categorization made by Watson, et al, (2013), in discussing the findings of the research.

Figure 2 reveals that twenty (20) studies used vertical integration which implies therefore, that most scholars used vertical approach in integrating sustainability. These included: 1, 2, 3, 8, 9, 10, 11, 12, 13, 14, 15, 16, 17, 19, 20, 25, 26, 27, 28, and 30. This finding agrees with Mulder, et al, (2000) results who stressed that the early SD courses were often a series of lectures added on to the existing programmes. This is also confirmed by Huntzinger, et al, (2007), in their analysis of Universities that integrate sustainability in the US and Canada. They reported that 'most of the examined universities bolted-on… sustainability in to their existing programmes' (p. 225). Bascoul, et al, (2013), in their study 'using an experiential business game to stimulate sustainable thinking in marketing education' found out that vertical integration through role playing games helps students to begin to reconstruct Life Cycle Analysis (LCA) of a product according to their own perceptions and, subsequently, to confront these perceptions with reality. The use of online reflective journal is another approach to vertical integration used by Betrabet Gulwadi (2009). She emphasized the success of the approach in education and nursing, and helps to foster critical thinking and engage learners in

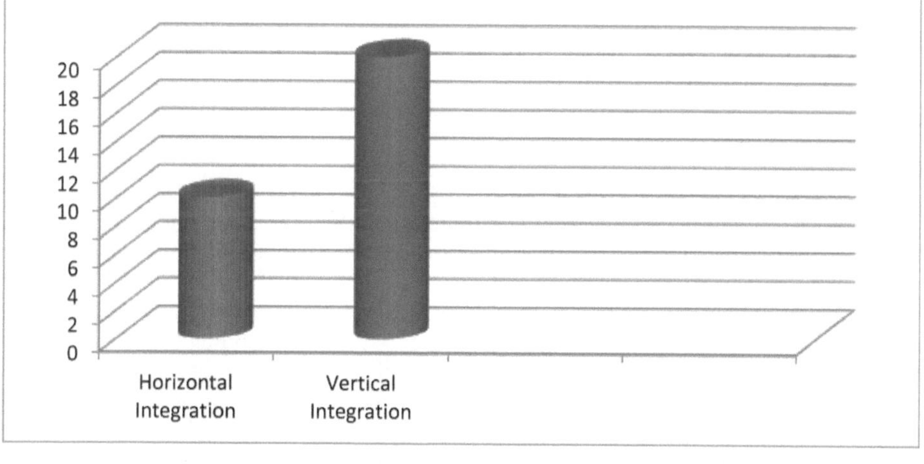

Figure 2. Vertical and Horizontal Integration of Approaches of Integrating Sustainability in HEIs

dialogue with theoretical concepts and to accept the applicability of concepts to real-world situations. The strength of using vertical integration as suggested by Lambrechts, et al., (2013) include that SD is explicitly stated in the competence matrix, as such encouraging integration in the curriculum. Another merit of vertical integration has to do with disseminating of a fresh course with SD contents as necessary for educating students about fundamental concepts and principles related to sustainability (Peet and Mulder, 2004, in Barrella and Watson, 2016).

However, Rose, Ryan and Desha (2015) warned that the danger of utilizing vertical integration is that it may not provide an adequate counter to 'unsustainability' reflected in other courses in the program. Also, the isolated nature of the sustainability content will not enable students to incorporate it into their professional practice. Furthermore, Peet and Mulder (2004), as cited by Barrella and Watson (2016), stressed that vertical integration alone may be insufficient because only teaching students about sustainability separate from core engineering concepts does not encourage them to incorporate sustainability into their professional designs and practices. Furthermore, Lambrechts, et al., (2013), maintained that using vertical integration implies that SD is an 'extra' topic, added to the matrix and clearly separated from other discipline-related competencies. Teaching sustainability separate from core engineering fundamentals, then students will view sustainability as an afterthought during the design process.

Figure 2 reveals that only ten (10) studies used horizontal integration (4, 5, 6, 7, 18, 21, 22, 23, 24 and 29). This finding confirmed Vargas and Maclean's (2015) assertion that although there have been many instances of SD integration in higher education engineering curricula, there has been little of strategic and systemic integration. They also affirmed that the vertical integration of SD through what they referred to as 'bolt-on' approach seem to be becoming extinct, and therefore used a horizontal integration in their study on minor programme on sustainability for the engineering curriculum at the University of Chile. Segalas, Ferrer-Balas and Mulder (2009), citing Holmberg et al., (2008) opined that embedding sustainability within the curriculum does not only mean including new contents. For engineers to continue to give their positive quota towards sustainability, SD ought to be a part of their paradigm and must impact their lifelong thinking. For this to occur, Sterling (2001), stressed that SD must becomes an integral part of engineering education programmes, not a mere 'add-on' to the 'core' parts of the curriculum.

Barrella and Watson (2016) also conducted a study to evaluate the efficacy of horizontal versus vertical integration of sustainability content into undergraduate engineering curricula, and found that students participating in a horizontally-integrated curriculum demonstrated more balanced understanding of sustainability, including the social dimension, as compared to students from a vertically-integrated program. Despite this fact as revealed by Barrella and Watson, horizontal integration of integrating SD is the most difficult strategy to implement. Inclusion of sustainability in all the competencies is the strength of horizontal integration, and is a larger framework for the competence matrix (Lambrechts, et al., 2013). However, Lambrechts, et al., (2013) stated that horizontal integration suffers from a drawback which include implicit integration of SD can make it into an 'optional' element; and integration in the curriculum is not guaranteed.

6 CONCLUSION

This study investigated approaches to integrating sustainability in higher education with special reference to undergraduate curricula, and the leading and predominant approaches identified. This is so important to higher education institutions and faculties who have not yet embed sustainability in their curricula, but have an intention to do so, especially in sub-Saharan Africa. This can help in having an insight of what approach might yield good results in their sustainability programmes. The authors classified the diverse approaches used in to two (2) broad groups of Vertical and Horizontal integrations. The findings of this study are in conformity with previously published suggestions that horizontal integration of SD is important for student's critical thinking, systemic thinking and problem solving.

REFERENCES

Al-Hagla, K. S. (2012). The role of the design studio in shaping an Architectural Education for sustainable development: The case of Beirut Arab University. *International Journal of Architectural Research: ArchNet-IJAR, 6*(1), 23-41.

Anand, C. K., Bisaillon, V., Webster, A., & Amor, B. (2015). Integration of sustainable development in higher education–a regional initiative in Quebec (Canada). *Journal of Cleaner Production, 108*, 916–923.

Annan-Diab, F., & Molinari, C. (2017). Interdisciplinarity: Practical approach to advancing education for sustainability and for the Sustainable Development Goals. *The International Journal of Management Education, 15*(2), 73–83. https://doi.org/10.1016/J.IJME.2017.03.006

Barrella, E. M., & Watson, M. K. (2016). Comparing the outcomes of horizontal and vertical integration of sustainability content into engineering curricula using concept maps *New developments in engineering education for sustainable development* (pp. 1-13): Springer.

Bascoul, G., Schmitt, J., Rasolofoarison, D., Chamberlain, L., & Lee, N. (2013). Using an experiential business game to stimulate sustainable thinking in marketing education. *Journal of Marketing Education, 35*(2), 168-180.

Betrabet Gulwadi, G. (2009). Using reflective journals in a sustainable design studio. *International Journal of Sustainability in Higher Education, 10*(1), 43-53.

Bussemaker, M., Trokanas, N., & Cecelja, F. (2017). An ontological approach to chemical engineering curriculum development. *Computers and Chemical Engineering, 106*, 927–941. https://doi.org/10.1016/j.compchemeng.2017.02.021

Cheah, S.-M., Yang, K., & Sale, D. (2012). *Pedagogical Approach to Integrate Sustainable Development into Engineering Curriculum*. Paper presented at the Proceedings of the 8th International CDIO Conference.

Doppelt, B. (2012). *The Power of Sustainable Thinking:" How to Create a Positive Future for the Climate, the Planet, Your Organization and Your Life"*: Routledge.

Du, X. Y., Su, L. Y., & Liu, J. L. (2013). Developing sustainability curricula using the PBL method in a Chinese context. *Journal of Cleaner Production, 61*, 80-88. doi: 10.1016/j.jclepro.2013.01.012

Dubois, R., Balgobin, K., Gomani, M. S., Kelemba, J. K., Konayuma, G. S., Phiri, M. L., & Simiyu, J. W. (2010). Integrating Sustainable Development in Technical and Vocational Education and Training.

Edvardsson Björnberg, K., Skogh, I.-B., & Strömberg, E. (2015). Integrating social sustainability in engineering education at the KTH Royal Institute of Technology. *International Journal of Sustainability in Higher Education, 16*(5), 639-649.

Garland, N. P., Khan, Z. A., & Parkinson, B. (2011). *Integrating Social Factors Through Design Analysis*. Paper presented at the DS 69: Proceedings of E&PDE 2011, the 13th International Conference on Engineering and Product Design Education, London UK, 08.-09.09. 2011.

Guerra, A. (2017). Integration of sustainability in engineering education: Why is PBL an answer? *International Journal of Sustainability in Higher Education, 18*(3), 436-454.

Halbe, J., Adamowski, J., & Pahl-Wostl, C. (2015). The role of paradigms in engineering practice and education for sustainable development. *Journal of Cleaner Production, 106*, 272-282. doi: 10.1016/j.jclepro.2015.01.093

Hayden, N. J., Rizzo, D. M., Dewoolkar, M. M., Neumann, M. D., Lathem, S., & Sadek, A. (2011). Incorporating a Systems Approach into Civil and Environmental Engineering Curricula: Effect on Course Redesign, and Student and Faculty Attitudes. *Advances in Engineering Education, 2*(4), n4.

Huntzinger, D. N., Hutchins, M. J., Gierke, J. S., & Sutherland, J. W. (2007). Enabling sustainable thinking in undergraduate engineering education. *International Journal of Engineering Education, 23*(2), 218.

Husgafvel, R., Martikka, M., Egas, A., Ribeiro, N., & Dahl, O. (2017). Development of A Study Module on and Pedagogical Approaches to Industrial Environmental Engineering and Sustainability in Mozambique. *International Journal of Higher Education, 6*(2), 50.

Iyer-Raniga, U., & Andamon, M. M. (2016). Transformative learning: innovating sustainability education in built environment. *International Journal of Sustainability in Higher Education, 17*(1), 105-122. doi: 10.1108/ijshe-09-2014-0121

Kapitulčinová, D., AtKisson, A., Perdue, J., & Will, M. (2018). Towards integrated sustainability in higher education–Mapping the use of the Accelerator toolset in all dimensions of university practice. *Journal of Cleaner Production, 172*, 4367-4382.

Khalifa, M. A., & Sandholz, S. (2012). Breaking Barriers and Building Bridges through Networks: An Innovative Educational Approach for Sustainability. *International Journal of Environmental and Science Education, 7*(2), 343-360.

Lambrechts, W., Mulà, I., Ceulemans, K., Molderez, I., & Gaeremynck, V. (2013). The integration of competences for sustainable development in higher education: an analysis of bachelor programs in management. *Journal of Cleaner Production, 48*, 65-73.
Leal Filho, W., Shiel, C., & Paço, A. (2016). Implementing and operationalising integrative approaches to sustainability in higher education: the role of project-oriented learning. *Journal of Cleaner Production, 133*, 126–135. https://doi.org/10.1016/j.jclepro.2016.05.079
Lozano, F. J., & Lozano, R. (2014). Developing the curriculum for a new Bachelor's degree in Engineering for Sustainable Development. *Journal of Cleaner Production, 64*, 136–146. https://doi.org/10.1016/j.jclepro.2013.08.022
Schmidt, M., Dr-Ing, D. (2016). Partnership between Technical Universities for Promoting Knowledge about Sustainability Standards within the Higher Education Curriculum. *Natsional'nyi Hirnychyi Universytet. Naukovyi Visnyk*(2), 152.
Martin S., (2008) Sustainable Development, Systems Thinking and Professional Practice", Journal of Education for Sustainable Development, 2:1, 2008; pp.31-40.
McConville, J. R., Rauch, S., Helgegren, I., & Kain, J.-H. (2017). Using role-playing games to broaden engineering education. *International Journal of Sustainability in Higher Education, 18*(4), 594-607.
Miñano, R., Fernández Aller, C., Anguera, Á., & Portillo, E. (2015). Introducing ethical, social and environmental issues in ICT engineering degrees. *Journal of Technology and Science Education, 5*(4).
Mintz, K., & Tal, T. (2013). Education for sustainability in higher education: a multiple-case study of three courses. *Journal of Biological Education, 47*(3), 140-149. doi: 10.1080/00219266.2013.821353
Ortega-Sánchez, M., Moñino, A., Bergillos, R. J., Magaña, P., Clavero, M., Díez-Minguito, M., & Baquerizo, A. (2018). Confronting learning challenges in the field of maritime and coastal engineering: Towards an educational methodology for sustainable development. *Journal of Cleaner Production, 171*, 733–742. https://doi.org/10.1016/j.jclepro.2017.10.049
Pérez-Foguet, A., Lazzarini, B., Giné, R., Velo, E., Boni, A., Sierra, M., ... Trimingham, R. (2017). Promoting sustainable human development in engineering: Assessment of online courses within continuing professional development strategies. *Journal of Cleaner Production, 172*, 4286–4302. https://doi.org/10.1016/j.jclepro.2017.06.244
Rhee, J., Oyamot, C. M., Speer, L., Parent, D. W., Basu, A., & Gerston, L. N. (2014). A Case Study of a Co-Instructed Multidisciplinary Senior Capstone Project in Sustainability. *Advances in Engineering Education, 4*(2).
Rose, G., Ryan, K., & Desha, C. (2015). Implementing a holistic process for embedding sustainability: A case study in first year engineering, Monash University, Australia. *Journal of Cleaner Production, 106*, 229–238. https://doi.org/10.1016/j.jclepro.2015.02.066
Rose, G., Ryan, K., & Desha, C. (2015). Implementing a holistic process for embedding sustainability: a case study in first year engineering, Monash University, Australia. *Journal of Cleaner Production, 106*, 229-238.
Rusinko, C.A., (2010). Integrating sustainability in higher education: a generic matrix. Int. J. Sustain. High. Educ. 11, 250e259
Segalàs, J., Ferrer-Balas, D., & Mulder, K. (2009). *Introducing sustainable development in engineering education: competences, pedagogy and curriculum.* Paper presented at the Proc. of the 37 th Annual Conference of the Society for Engineering Education (SEFI), Rotterdam, The Netherlands.
Segalas, J., Ferrer-Balas, D., & Mulder, K. F. (2010). What do engineering students learn in sustainability courses? The effect of the pedagogical approach. *Journal of Cleaner Production, 18*(3), 275-284.
Shing, C. K., Mohamad, Z. F., & Raman, A. A. A. (2016). Integrating Components of Sustainability into Chemical Engineering Curricula. *International Journal of Engineering Education, 32*(6), 2653-2664.
Siddiqui, M. K., et al. (2012). "Integrating sustainability in the curriculum through capstone projects: A case study." Proc., American Society for Engineering Education Annual Conf. and Exposition, American Society for Engineering Education, Washington, DC.
Staniškis, J. K., & Katiliute, E. (2016). Complex evaluation of sustainability in engineering education: Case & analysis. *Journal of Cleaner Production, 120*, 13–20. https://doi.org/10.1016/j.jclepro.2015.09.086
Tejedor, G., Segalàs, J., & Rosas-Casals, M. (2018). Transdisciplinarity in higher education for sustainability: How discourses are approached in engineering education. *Journal of Cleaner Production, 175*, 29–37. https://doi.org/10.1016/j.jclepro.2017.11.085
Thürer, M., Tomašević, I., Stevenson, M., Qu, T., & Huisingh, D. (2017). A Systematic Review of the Literature on Integrating Sustainability into Engineering Curricula. *Journal of Cleaner Production.*
UNESCO, (2002). Education for Sustainability. From Rio to Johannesburg: Lessons learnt from a Decade of Commitment. World Summit on Sustainable Development, (UNESCO). Johannesburg. Available at http://unesdoc.unesco.org/images/0012/001271/127100e.pdf

Vargas, L. S., & Mac Lean, C. (2014). A Minor Programme on Sustainability for the Engineering Curriculum at the University of Chile *Integrating Sustainability Thinking in Science and Engineering Curricula* (pp. 21-29): Springer

Von Blottnitz, H., Case, J. M., & Fraser, D. M. (2015). Sustainable development at the core of undergraduate engineering curriculum reform: A new introductory course in chemical engineering. *Journal of Cleaner Production, 106*, 300–307. https://doi.org/10.1016/j.jclepro.2015.01.063

Watson, M. K., Lozano, R., Noyes, C., & Rodgers, M. (2013). Assessing curricula contribution to sustainability more holistically: Experiences from the integration of curricula assessment and students' perceptions at the Georgia Institute of Technology. *Journal of Cleaner Production, 61*, 106-116.

Willamo, R., Helenius, L., Holmström, C., Haapanen, L., Sandström, V., Huotari, E., ... Kolehmainen, L. (2018). Learning how to understand complexity and deal with sustainability challenges – A framework for a comprehensive approach and its application in university education. *Ecological Modelling, 370*, 1–13. https://doi.org/10.1016/j.ecolmodel.2017.12.011

APPENDIX

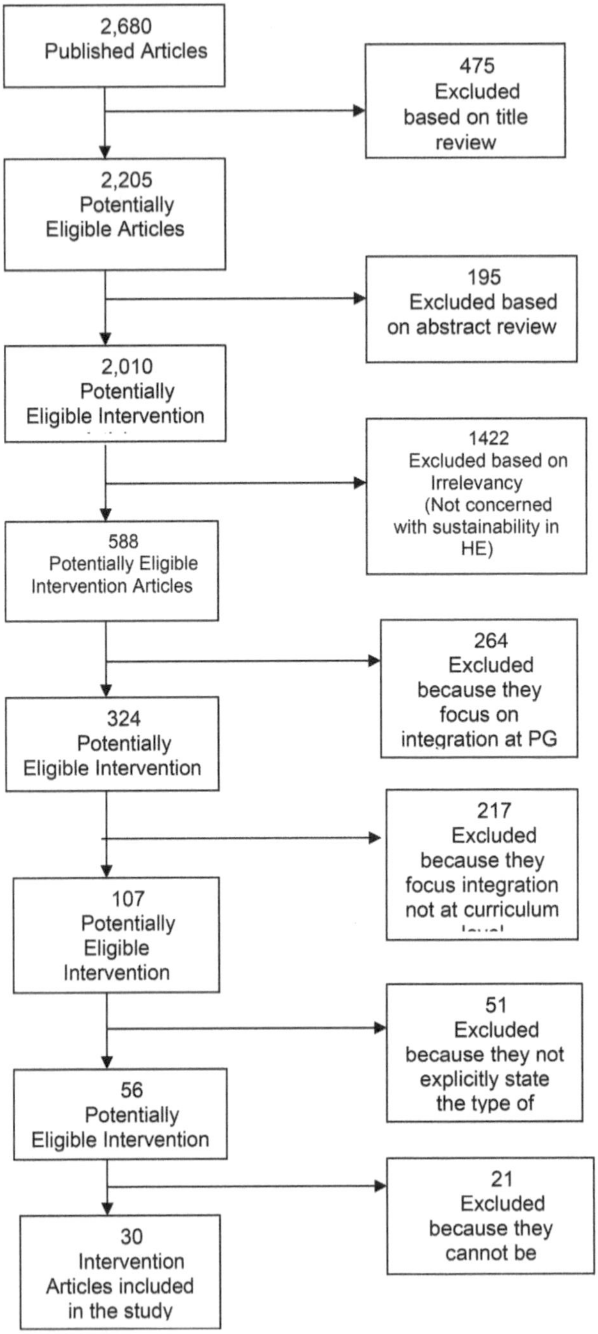
2,680
Published Articles
475
Excluded
based on title
review
2,205
Potentially
Eligible Articles
195
Excluded based
on abstract review
2,010
Potentially
Eligible Intervention
1422
Excluded based on
Irrelevancy
(Not concerned
with sustainability in
HE)
588
Potentially Eligible
Intervention Articles
264
Excluded
because they
focus on
integration at PG
324
Potentially
Eligible Intervention
217
Excluded
because they
focus integration
not at curriculum
107
Potentially
Eligible
Intervention
51
Excluded
because they not
explicitly state
the type of
56
Potentially
Eligible Intervention
21
Excluded
because they
cannot be
30
Intervention
Articles included
in the study

TVET Towards Industrial Revolution 4.0– Hazirah Noh@Seth et al. (eds)
© 2020 Taylor & Francis Group, London, ISBN 978-0-367-24273-2

Factors that influence the effectiveness of the teaching and learning of electronic subjects in vocational colleges

A.N.M. Nasir, A. Ahmad, M.F. Ahmad, N.H.A. Wahid & N. Suhairom
Department of Technical and Engineering Education, School Of Education, Faculty of Social Science and Humanities, Universiti Teknologi Malaysia, Johor, Malaysia

ABSTRACT: Practical teaching is an approach that is used in teaching and learning in vocational colleges in Malaysia. The effectiveness of a good practical teaching approach not only depends on the lecturers' competence in carrying out teaching and learning activities, but is also influenced by other factors. Hence, this paper will look at what factors influence the teaching and learning of practical electronic subjects in vocational colleges. The researcher uses a descriptive research approach and uses a questionnaire as a research instrument to obtain data. A total of 135 lecturers and 238 students from 58 vocational colleges that offer Electrical and Electronics Engineering Technology courses were selected in this research. Descriptive data analysis has been used in this study, as the mean and standard deviation is obtained. The findings show that the workshop environment and the use of technology have a positive effect on the effectiveness of teaching and learning sessions in vocational colleges. The findings of this study will help teachers to ensure that factors such as the workshop environment and the use of technology are given the correct emphasis when planning lessons and practical learning in the classroom.

1 INTRODUCTION

The survival of the teaching and learning process is influenced by certain factors. These factors are indicators and things that need to be addressed in order that the ideal teaching and learning process can be achieved. Lecturer's teaching is a key aspect of the initial process of the teaching process in a workshop, but some factors need to be taken into account in ensuring that it progresses smoothly. The workshop environment and the use of technology during the teaching process are among the factors that need to be addressed (Royo & Zainun, 2005; Yahaya & Sharudin, 2008; Felder & Brent, 2012).

Lecturers' teaching in the classroom is a process of knowledge delivery that requires an ideal situation in order to ensure that knowledge is effectively delivered to the students, while the learning process of the students in the classroom is a process of acceptance of knowledge from the lecturers who are teaching them (Nasir et al., 2018a; Nasir et al., 2018b). However, the process of teaching and learning is also influenced by some external factors that can influence a smooth teaching and learning process. Lecturers and students should look at these factors when preparing for teaching and learning in the classroom. It involves factors such as workshops and the use of technology.

1.1 *Workshop Environment*

The majority of the teaching and learning time for practical subjects is usually undertaken in the workshop. The workshop environment is one of the factors in the teaching and learning process that is either good or it is not. This is because if the workshop environment is clean, cheerful, organized, and bright it influences the smoothness of the learning process and the achievement of students (Meng, 1994; Trembley et al., 1999; Yahaya & Sharudin, 2008; Azizi et al., 2011).

Good and smooth air flow affects teaching and learning. This is because the practical work in the workshop will cause a high level of heat output among lecturers and students (Mustapha, 2000). They will receive less oxygen and feel tired and too lethargic to continue focusing on the workshop. Good ventilation in the workshop is required to be not only clean, but also free from impurities such as hazardous dust. Students who are in a comfortable state can give their maximum concentration to the practical learning process in the workshop (Long, 1982; Royo & Zainun, 2005; Yahaya & Sharudin, 2008). Convenient work comforts provide a positive impact, not only to help prevent accidents, but also to improve efficiency and focus when undertaking the work (Nor, 2000).

Practical equipment is usually placed in a designated place. The orderly arrangement of this practical equipment will facilitate lecturers and students to retrieve and return it (Mustapha, 2000). Practical equipment should be used according to the requirements of the practical syllabus so that it is easy to find and reinstate (Khalid, 2002). In addition, practical equipment should also be maintained, repaired, stored correctly and used properly to avoid damaging it. Each piece of practical equipment in the workshop has a certain lifespan. Practical equipment that works well with no damage can accelerate the practical teaching process (Khalid, 2002). The use of practical equipment should be planned before the teaching session so that the practical work will proceed smoothly and well (Saleh, 2002).

1.2 *Technology Use*

The use of computers as a teaching aid is one of the aspects emphasized in the process of improving the quality of teaching. The use of technology in practical teaching can have a significant impact through the app visualization of a practical application system, its use in the industry and its basic concepts. However, teaching processes based on this computer technology are not always able to run smoothly due to the lack of proficiency of the lecturer in using the technology (Sani et al., 2004; Muslim et al., 2006; Zakaria et al., 2008).

The use of technology in teaching gives lecturers more space to make things more creative and facilitates students to understand the content of the teaching. The combination of using both technology and practicality in teaching ensures that students are more focused and find it easier to understand the teaching content in the workshop. This is in line with the study of Gani et al. (2006), who suggested that lecturers who are knowledgeable in computer use during the teaching session can have a positive impact on the smoothness of the teaching and learning process in the classroom.

The use of technology in the learning process is one of the more widely used methods nowadays. It is used as a teaching and learning aid in the classroom, such as viewing videos that are relevant to the topic being studied. The use of technology in practical teaching is a way of providing lecturers with teaching aids in the workshop. Instructional materials are a factor that can influence student learning, as they are tools that facilitate the students to understand the content in a more interesting and precise way within the classroom (Azizi et al., 2011).

2 RESEARCH METHODOLOGY

The researcher uses a descriptive research approach and uses a questionnaire as a research instrument to obtain data. This research used a random cluster sampling technique. A total of 135 lecturers and 238 students from 58 vocational colleges that offer Electrical and Electronics Engineering Technology courses were selected for this research (Bahagian Pendidikan Teknik dan Vokasional, 2017). The survey is focused on the factors that influence the effectiveness of teaching and learning on

2.1 *Instrument Validity And Reliability*

Researchers submitted a questionnaire that had been created to two experts in electronic teaching lessons for the purpose of reviewing the content. After the questionnaire was validated by the experts, the researcher conducted a pilot study by giving the questionnaire to 30 first-year students on Electronic Technology courses and thirty (30) lecturers of electronics at vocational colleges to test the reliability of the instrument. In the pilot study, the minimum number of respondents involved were thirty (Chua Yan Piaw, 2014). The appropriate reliability coefficient to be used in the assessment of this study is that it has a greater coefficient of reliability than 0.60 (Ghafar, 2003). The results for the reliability test are shown in Table 1.

Table 1. Reliability score.

Elements	Reliability score
Factors (Lecturers as respondents)	0.897
Factors (Students as respondents)	0.891

2.2 *Factor Scale Indicators*

In this study, the selection of factor scales that influence the practical teaching methods in the vocational colleges is divided into two, namely: *Influence* and *Not an influence.* Table 2 shows the level used by the researcher, which is adapted from Chua Yan Piaw (2014).

Table 2. The scale of factors influencing practical teaching methods in the vocational colleges.

Mean score	Description
3.51 – 6.00	Influence
1.00 – 3.50	Not an influence

3 RESULT

The factors studied were the workshop environment and the use of technology in teaching and learning. Tables 3 and 4 show the mean of the factors affecting the teaching and learning of electronic subjects in the vocational colleges.

Table 3. Factors affecting the teaching of electronic subjects in the vocational colleges.

Factors	Mean	SD	Description
Workshop environment	5.38	0.59	Influence
Use of technology	5.04	0.57	Influence

Table 4. Factors affecting the learning of electronic subjects in the vocational colleges.

Factors	Mean	SD	Description
Workshop environment	5.29	0.67	Influence
Use of technology	5.05	0.70	Influence

Based on the research data, it can be seen that the workshop environment and the use of technology influence the teaching of practical electronic subjects in the vocational colleges for lecturers of electronics.

Based on the research data, it can be seen that the workshop environment and the use of technology influenced the students' learning of electronic subjects in the vocational colleges.

4 DISCUSSION

Based on the findings, the mean score shows that the workshop environment is able to influence teaching and learning in the classroom during practical teaching sessions. This is in line with the studies by Royo and Zainun (2005) and Yahaya and Sharudin (2008), which explain that the workshop environment should have a conducive space that is comfortable in order to provide the ideal environment for carrying out the practical teaching process. Good and systematic workshop management can play a large role in the effectiveness of teaching and learning in the workshop. The cleanliness and orderly arrangement of the workspace leads to the student having the ability to do practical work in comfort and provides a conducive working environment (Yahaya & Sharudin, 2008).

If the lecturers and students are comfortable in the workshop then they can be more focused during practical teaching sessions. A poorly managed workshop environment can create a variety of unexpected accidents and incidents. It not only affects the teaching process, but also causes injuries to students (Yahaya & Sharudin, 2008; Azizi et al., 2011). Convenience in the workshop during practical teaching sessions includes such things as seating arrangements and desks for experiments, good lighting effects, good ventilation, good organization, and complete and non-damaged practical arrangements. Convenient work comforts provide a positive impact, not only to help prevent accidents, but also to improve efficiency and focus when undertaking work (Nor, 2000).

Practical desk arrangements need to be quite conducive, including layout, hygiene, and even the number of students to each experimental desk used. This comfort will give students space and a venue for more active students during practical teaching sessions. The number of students who are overweight can cause the control within the group to not work well. A good workshop environment should have a good lighting and ventilation system. Workshops with less light and poor airflow may result in undesirable incidents, such as accidents and less oxygen, and can even cause students to not focus in workshops (Mustapha, 2000; Yahaya & Sharudin, 2008). Having enough lighting in the workshop allows lecturers to monitor the movements of the students and the situations experienced by them during their practical teaching sessions. This can give lecturers indirect feedback regarding the students' concentration and the process in practical teaching sessions. Consequently, good airflow in practical workshops can provide enough oxygen to the students. It is important to prevent students from becoming drowsy and give them more focus during practical teaching.

Practical teaching and learning in the workshop is also influenced by the completeness of good practical equipment. Practical equipment that works well and which is not damaged can accelerate and launch the practical teaching process (Khalid, 2002). Corrupt and insufficient equipment can cause the teaching process to become disrupted. It will take time out of the lecturer's practical lessons if they are forced to search for new tools or if the tools that are used are inadequate. This can result in the students being disconnected and causing them to drop out of practical lessons (Saleh, 2002).

The arrangement of practical equipment in an organized and neat workshop will facilitate lecturers and students to take and use them if necessary. They should be placed in an easy-to-see place and be located in an easy-to-reach area during practical teaching sessions (Mustapha, 2000). This can reduce time spent on the practical preparation session, thus facilitating them to focus when the practical teaching session takes place.

The use of technology by lecturers during teaching sessions can also influence teaching and learning. The use of computers as a teaching tool can also help to facilitate practical teaching (Sani et al., 2004; Muslim et al., 2006; Zakaria et al., 2008). New information and information in line with the industry's needs can be obtained through internet searches. Lecturers with new information can provide different ideas and insights into the field of industry if they are exposed to new knowledge.

Lecturers also need to be responsible for assigning students to seek new knowledge, as well as teaching in current workshops and industries. This can be done either before the teaching session to give the students an idea of the topic or after a practical teaching session is conducted to reinforce new knowledge and their understanding of the practical teaching. Providing teaching using a computer with slideshow use conducted by students can give effect to teaching with more focus (Long, 1980; Azizi et al., 2011). Students will focus more when exposed in front of the class. In addition, lecturers and students agree that the use of the speaker system can help the teaching and learning process in electronic workshops. If lecturers use a clear voice it can make it easier for students to understand the contents of the teaching. Appropriate approaches used by lecturers during practical teaching sessions, when assisted by factors in the workshop, have a positive impact on the teaching and learning process of electronic practice at vocational colleges.

5 CONCLUSION

Factors such as the workshop environment and the use of technology have an impact on ensuring that practical teaching and learning at vocational colleges works well. A conducive workshop environment, complete and industry-specific equipment and a good air flow are catalysts for ensuring added value for students and lecturers. Combined with the use of technology in teaching and learning, this can ensure that current and accurate information can be communicated well to students.

REFERENCES

Azizi, N.E., Maalip, H. & Yahaya, N. (2011). Perbandingan faktor yang mempengaruhi proses pengajaran dan pembelajaran antara sekolah di bandar dan luar bandar. *Journal of Educational Management, 1*(3), 1–17.

Bahagian Pendidikan Teknik dan Vokasional. (2017). *Data Populasi Guru dan Pelajar Kolej Vokasional.*

Chua Yan Piaw (2014). Kaedah Statistik Penyelidikan, Buku 2. Asas Statistik Penyelidikan (Edisi Ketiga). Mc Graw Hill, Selangor

Felder, R.M. & Brent, R.E. (2012). *Effective teaching: A workshop.* West Lafayette, IN: Purdue University. Retrieved from https://engineering.purdue.edu/Engr/AboutUs/Administration/AcademicAffairs/Research/Research/Resources/Teaching/effective-teaching.pdf

Gani, A.W.I., Siarap, K. & Mustafa, H. (2006). Penggunaan Komputer Dalam Pengajaran-Pembelajaran Dalam Kalangan Guru Sekolah Menengah: Satu Kajian Kes Di Pulau Pinang. *Kajian Malaysia, 24*(1).

Ghafar, M.N.A. (2003). *Reka bentuk tinjauan soal selidik pendidikan.* Skudai, Malaysia: Penerbit UTM.

Khalid, A. (2002). *Amalan Pengurusan Bengkel di Sekolah Menengah Vokasional yang Telah di Naik Taraf. Satu Tinjauan* (Bachelor's thesis, Universiti Teknologi Malaysia, Johor, Malaysia).

Long, A. (1980). *Psikologi Pendidikan.* Kuala Lumpur, Malaysia: Dewan Bahasa Dan Pustaka.

Long, A. (1982). *Pedagogi Kaedah Am Mengajar*. Petaling Jaya, Malaysia: Amiza Publising Sdn Bhd.

Meng, E.A. (1994). *Psikologi Dalam Bilik Darjah*. Kuala Lumpur, Malaysia: Fajar Bakti.

Muslim, N.A., Yahya, N. & Abidin, N. (2006). *Persepsi Guru-Guru Terhadap Penggunaan Teknologi Maklumat dan Komunikasi*. Working paper for SITMA Seminar Expansion Seminar, Terengganu, 19–20 August.

Mustapha, H. (2000). *Amalan Peraturan Keselamatan Bengkel di Kalangan Pelajar 4 STP (Kejuruteraan Awam/Jentera/Elektrik/Kemahiran Hidup) di Fakulti Pendidikan, UTM, Skudai: Satu tinjauan* (Bachelor's thesis, Universiti Teknologi Malaysia, Johor, Malaysia).

Nasir, A.N.M., Ahmad, A., Udin, A., Wahid, N.H.A. & Noordin, M.K. (2018a). Competency level of technical knowledge for electronic teachers in vocational college, Malaysia. *Advanced Science Letters*, *24*(4), 2796–2798.

Nasir, A.N.M., Seth, N.H.N., Noordin, M.K., Suhairom, N. & Ahmad, A. (2018b). Technical teaching competency of the electronic teachers in vocational college, Malaysia. *The Turkish Online Journal of Design, Art and Communication – TOJDAC, September 2018 Special Edition*, 1693–1696.

Nor, M.N.M. (2000). *Amalan keselamatan bengkel di kalangan pelajar kursus amalan bengkel mesin di Sekolah Menengah Teknik Kemaman, Terengganu: Satu tinjauan* (Bachelor's thesis, Universiti Teknologi Malaysia, Johor, Malaysia).

Royo, M.A. & Zainun, Z.A.B. (2005). *Faktor-Faktor Yang Mempengaruhi Kualiti Keputusan Peperiksaan Dalam Mata Pelajaran Lukisan Kejuruteraan Di Sekolah Menengah Teknik Perdagangan Johor Bahru*. Johor, Malaysia: Fakulti Pendidikan, Universiti Teknologi Malaysia.

Saleh, R. (2002). *Persepsi Pelajar Terhadap Amalan Keselamatan Semasa Melakukan Kerja-kerja Amali di Dalam Bengkel Automotif* (Bachelor's thesis, Universiti Teknologi Malaysia, Johor, Malaysia).

Sani, M., Nordin, M. & Roslee, M. (2004). Budaya ICT di Kalangan Guru-Guru Sains dan Matematik: Trend dan Isu. *Jurnal Pendidikan*, *7*, 15–24.

Trembley, S., Ross, N. & Berthelot, J.M. (1999).The relationship between environmental quality of school facilities and student performance. *Environmental International*, *12*, 147–159.

Yahaya, A. & Sharudin, S.A. (2008). *Faktor Yang Mempengaruhi Keberkesanan Pengajaran Dan Pelajaran Di Dalam Bengkel Vokasional Di Dua Buah Sekolah Menengah Teknik Di Negeri Sembilan*. Johor, Malaysia: Fakulti Pendidikan, Universiti Teknologi Malaysia.

Zakaria, M.A.Z.M., Aris, B. & Harun, J. (2008). *Kemahiran ICT di kalangan guru-guru pelatih UTM: Satu tinjauan*. 1st International Malaysian Educational Technology Convention, Kuala Lumpur, August.

TVET Towards Industrial Revolution 4.0– Hazirah Noh@Seth et al. (eds)
© 2020 Taylor & Francis Group, London, ISBN 978-0-367-24273-2

Author Index